T0306175

# FRAM: THE FUNCTIONAL RESONANCE ANALYSIS METHOD

# FRAM: The Functional Resonance Analysis Method

### Modelling Complex Socio-technical Systems

ERIK HOLLNAGEL

*University of Southern Denmark, Denmark*

## CRC Press

Taylor & Francis Group

Boca Raton  London  New York

CRC Press is an imprint of the
Taylor & Francis Group, an **informa** business

CRC Press
Taylor & Francis Group
6000 Broken Sound Parkway NW, Suite 300
Boca Raton, FL 33487-2742

© 2012 by Erik Hollnagel
CRC Press is an imprint of Taylor & Francis Group, an Informa business

No claim to original U.S. Government works

Printed on acid-free paper
Version Date: 20160226

International Standard Book Number-13: 978-1-4094-4551-7 (Paperback) 978-1-4094-4552-4 (Hardback)

**Visit the Taylor & Francis Web site at**
**http://www.taylorandfrancis.com**

**and the CRC Press Web site at**
**http://www.crcpress.com**

# Contents

# List of Figures

# List of Tables

# Prologue

The Functional Resonance Analysis Method – in the following referred to as the FRAM – does in a way itself represent what it tries to describe, namely that sometimes things happen without clearly recognised causes. In contrast to many other safety assessment methods, such as the Fault Tree developed in 1962 to ensure the safety of the Minuteman Intercontinental Ballistic Missile or the bulk of the first-generation Human Reliability Assessment (HRA) methods developed in response to the accident at Three Mile Island in 1979, the FRAM seems to have emerged from a diffuse background – or from a confused mind.

The first coherent presentation of the basic principles of the FRAM was in E. Hollnagel (2004), *Barriers and Accident Prevention* (Farnham: Ashgate). The idea had, however, been around for several years, and been part of various courses given at the University of Linköping. In hindsight there were four main sources of inspiration, namely Floyd Allport's 1954 paper about event-structures, Margoroh Maruyama's 1963 paper about the Second Cybernetics, several descriptions of the Structured Analysis and Design Technique, and a notion that stochastic resonance might be a good idea, based on an article in *New Scientist* from 1996. To that must be added the ever-present and growing dissatisfaction with the way safety issues were addressed, especially how accidents and incidents were explained. This dissatisfaction, shared with many colleagues, itself became the motivation for the development of Resilience Engineering. But that is a different story.

The idea of the FRAM was well received from the very beginning, and the development of the method from the 2004 book until today would not have happened without the interest, questions and contributions of a great many people. Although it is impossible to produce a complete list, I will nevertheless take this opportunity to thank (in alphabetical order) Karen Cardiff, Pedro Ferreira, Ivonne Herrera, Akinori Komatsubara, Jörg Leonhardt, Luigi Macchi, Elaine Pelletier, Shawn Pruchnicki, Rob Robson, Gunilla Sundström, Sebastien Travadel, Camilla Tveiten, Rogier Woltjer – and last but not least Denis Besnard and the whole FRAMily. With many apologies to those that have not been named.

Erik Hollnagel

# Chapter 1

# The Need

Virtually all design is conducted in a state of relative ignorance of the full behaviour of the system being designed.

(Henry Petroski)

## A State of (Relative) Ignorance

When a system is designed there is from the very beginning a need to know how it will function. Indeed, the very reason for system design is to construct an artefact that provides an intended functionality. In Henry Petroski's book about 'Design Paradigms,' from which the above epigraph is taken, the topic was engineering design, and the majority of examples were physical structures and static systems, such as bridges. The 'behaviour' of a bridge is seemingly simple: it just has to be there and to maintain its own structure in order to allow safe passage for whoever or whatever uses it. Yet even in this case there is a 'relative ignorance' of what may possibly happen, as numerous examples of collapsing structures have demonstrated.

While design may not require perfect knowledge it at least requires acceptable ignorance. Partial ignorance is unavoidable in principle as well as in practice, but it should be so little that the consequences are unnoticeable. In other words, we must be reasonably sure that the systems we build will do what they are supposed to do, that they will function reliably as intended and that they additionally will not do anything they are not supposed to do. The latter, in the case of a bridge, means collapsing or falling down. Yet history is full of examples of just that happening from the Dee bridge disaster (1847) to the spectacular collapse of the Tacoma Narrows in 1940 and more recently to the failure of the I-35W Mississippi River bridge (2007).

## The Reasons for Ignorance

There are several reasons for this relative ignorance. One is that the properties of the materials may not be known well enough, neither how they will behave under extreme conditions (pressure, temperature, wind and so on) nor how they will behave over time (ageing, degradation). Another is that it is uncertain what external conditions the system will be subjected to. In the case of bridges, the conditions refer to weather, changes in the chemical composition of the air, quality of maintenance, changes in 'customer' characteristics (for example, heavier cars and more traffic) and so on. A third reason is that work usually is done under conditions of insufficient time, information and resources. The only thing we can know for certain is that things will happen that we have never thought of, although that in itself is of limited comfort.

If the situation is difficult for structures and static systems, it is even worse for dynamic systems. (Note, by the way, that so far we only have referred to nominally technical systems, for example, a bridge as a bridge. Even here, the influence of social systems is obvious, for instance in how well the technical system is maintained, how well the components meet the specifications or requirements, how well the system is designed and built and so on.) For dynamic systems it is necessary to consider also the state of the parts and components as well as the dependencies among them. Energy – and information – must be provided, mass and materials must be moved around, substances will be transformed and so on. This creates literally countless dependencies of which there is relative ignorance, but which nevertheless must work as planned in order for the system to fulfil its purpose. But as it is well-nigh impossible to foresee all possible combinations, even for purely technological systems, surprises abound.

## Ignorance, Risk and Safety

Petroski's lament can be extended from the design of technical systems to include accident investigation and risk assessment as well. Virtually all accident investigations and virtually all risk assessments are conducted in a state of relative ignorance of the full behaviour of the system being analysed – and in some cases even in a state of ignorance about the typical behaviour.

In relation to event investigations, we can rarely if ever get all the information we need about what happened. One reason is that the search for information is influenced by biases and practical constraints. One such bias is the What-You-Look-For-Is-What-You-Find (WYLFIWYF) principle, which means that we look for what we assume is important. This precludes us from finding anything that we do not look for – serendipity excepted. Another bias is 'illusory comprehension'. The fact that we can squeeze events into pre-existing explanatory frameworks, all of which imply causality, means that we see causality even though it may not really be there. Among the practical constraints is the all too frequent lack of time, which means that the search for information is stopped when an acceptable explanation has been found, even though this may be incomplete or incorrect.

In relation to risk assessment, one source of ignorance is the inescapable uncertainty about what the future will bring. An observation made by many philosophers, and often repeated by politicians, is that we cannot know with certainty what will happen in the future. The Danish philosopher Søren Kierkegaard (1813–1855) noted that while life can only be understood backwards, it must be lived forwards. Samuel Coleridge (1772–1834) somewhat more poetically noted that 'the light which experience gives us is a lantern on the stern which shines only on the waves behind us'. A second source of ignorance is that most of the models or representations we use are so oversimplified that their validity is questionable. In an event tree, for instance, it is assumed that the chosen representation principle (binary branching) is an acceptable representation of reality. But that is clearly not the case, both because a distinction between fail and succeed is relative rather than absolute, and because things rarely develop in the way that was expected. A third source is the lack of imagination that partly is innate, partly comes from familiarity and habituation. A textbook example of that is Alan Greenspan's characterisation of the 2008 financial crisis as a 'once-in-a-century credit tsunami … that … turned out to be much broader than anything I could have imagined'.

## Ignorance, Complexity and Intractability

Ignorance of the future (and to some extent also ignorance of the past) is sometimes attributed to the degree of complexity of the systems we are dealing with, or simply to the purported fact that 'today's systems are – or have become – complex'. Complexity is, however, not a well-defined concept, as the following definitions exemplify:

- Mathematical complexity is a measure of the number of possible states a system can take on, when there are too many elements and relationships to be understood in simple analytic or logical ways.

- Pragmatic complexity means that a description, or a system, has many variables.
- Dynamic complexity refers to situations where cause and effect are subtle, and where the effects over time of interventions are not obvious.
- Ontological complexity has no scientifically discoverable meaning, as it is impossible to refer to the complexity of a system independently of how it is described.
- Epistemological complexity can be defined as the number of parameters needed to describe a system fully in space and time. While epistemological aspects can be decomposed and interpreted recursively, ontological aspects cannot.

It might indeed be asked whether there is a state of relative ignorance because the systems we deal with are complex, or whether we call the systems complex because we do not have – and possibly cannot have – complete knowledge about them.

While we may entertain the hope that complete knowledge in principle is possible for technological systems (barring the vagaries of software), there is no reason for such optimism in the case of socio-technical systems. Here ignorance is a fact of life because it is impossible fully to define or describe the parameters in space or time even if we knew what they were. The main reason for this is, however, not that there are too many parameters, but rather that the systems are dynamic, that is, that they continuously change.

In order for a system to be understandable it is necessary to know what goes on 'inside' it, to have a sufficiently clear description or specification of the system and its functions. The same requirements must be met in order for a system to be analysed and in order for its risks to be assessed. That this must be so is obvious if we consider the opposite. If we do not have a clear description or specification of a system, and/or if we do not know what goes on 'inside' it, then it is clearly impossible effectively to understand it, and therefore also to investigate accidents or assess risks.

The presence of the (relative) ignorance clashes with the assumptions of established safety analysis methods. These assumptions are a heritage from the large-scale technological systems for which the first safety assessment methods were developed in the late 1950s. Although the underlying assumptions rarely are stated explicitly, they are easy to recognise by looking at established methods, such as FMEA (Failure Mode and Effects Analysis), HAZOP (Hazard and Operability Study), Fault Trees and so on. The four main assumptions are:

- A system can be decomposed into meaningful elements (parts or typically components). Similarly, events can be decomposed into individual steps or acts. (The principle of decomposition is, of course, in conflict with the holistic principle that the whole is more than the sum of the parts.)

- Parts and components will either work or fail. In the latter case, the probability of failure can be analysed and described for each part or component individually. This is part of the rationale for focusing on the human error probability, and indeed for classifications of human errors.
- The order or sequence of events is predetermined and fixed as described by the chosen representation. If a different sequence of events needs to be considered, it is necessary to produce a new version of the representation, for example, a new event tree or fault tree.
- Combinations of events are orderly and linear. They can be described by standard logical operators, and outputs are proportional to inputs.

Although these assumptions may be warranted for technological systems, it is highly questionable whether they apply to social systems and organisations, or to human activities. Models and methods that require that the system in focus can be fully described will for that reason not be suitable for socio-technical systems, neither for accident analysis and nor for risk assessment. It is therefore necessary to look for methods and approaches that can be used for systems that are incompletely described or underspecified. The two types of systems can be called tractable and intractable, respectively. The differences are summarised in Table 1.1 below.

**Table 1.1    Tractable and intractable systems**

|  | Tractable system | Intractable system |
|---|---|---|
| Number of details | Description are simple with few details | Description are elaborate with many details |
| Rate of change | Low; in particular, the system does not change while being described | High: the system changes before a description can be completed. |
| Comprehensibility | Principles of functioning are completely known | Principles of functioning are partly unknown |
| Characteristic of processes | Homogeneous and regular | Heterogeneous and possibly irregular |

The differences between tractable and intractable systems can be illustrated by two examples. First consider a tractable system, such as a car assembly line. Here descriptions are (relatively) simple with only a small number of details. Work is meticulously planned and scheduled so that the assembly can be as efficient as possible and produce cars of a high quality. The rate of change is low, and usually the result of a planned intervention. Work is dominated by routine and is therefore homogeneous and highly regular. Finally, the comprehensibility is high, meaning that there is little, if anything, that is not understood in detail. The system is therefore tractable, which means that it can be specified in great detail and that decomposition is a natural approach to understand it better.

Then consider an intractable system, such as an emergency room (ER) in a hospital – or for that matter an emergency management room anywhere. Descriptions of such systems are elaborate and with many details since work is non-routine and the same situation rarely occurs twice. Intractable systems are heterogeneous rather than homogeneous. The rate of change is high, which means that the system – and its performance – is irregular and possibly unstable. Unlike a car assembly plant, work in an ER is difficult to plan because it is impossible to know when patients will arrive, how many there will be and what kind of treatment they require. Finally, comprehensibility is low, because not everything is understood in detail. The system is therefore intractable, which means that it cannot be specified in detail, and that it does not make sense to decompose it.

## Systems Redefined

Systems are usually defined with reference to their structure, that is, in terms of their parts and how they are connected or put together. Common definitions emphasise both that the system is a whole, and that it is composed of independent parts or objects that are interrelated in one way or another. Definitions of this type make it natural to rely on the principle of decomposition to understand how a system functions, and to explain the overall functioning in terms of the functioning of the components or parts – keeping in mind, of course, that the whole is larger than the sum of the parts.

It is, however, entirely possible to define a system in a different way, namely in terms of how it functions rather than in terms of what the components are and how they are put together. From this perspective, a system is a set of coupled or mutually dependent functions. This means that the characteristic performance of the system – of the set of functions – cannot be understood unless it includes a description of all the functions, that is, the set as a whole. The delimitation of the system is thus not based on its structure or on relations among components (the system architecture). An organisation, for instance, should not be characterised by what it is but by what it does. Neither should it be characterised by the people who are in a given place (on the organisation chart or in reality) but by the functions they performs. One consequence of a functional perspective is that the distinction between a system and its environment, and thereby also the system boundary, becomes less important; see the discussion of foreground and background functions in Chapter 5.

## From Probability to Variability

One important development in the history of industrial safety was the transition that happened in response to the accident at the Three Mile Island nuclear power plant in 1979. This led to a change in focus from technology alone to human performance and the ways in which this could go wrong. Since established safety practices required that the probability of a failure or malfunction could be calculated, this spawned numerous proposals for how to calculate the probability of a 'human error'.

To cut a long story short, the so-called first generation of Human Reliability Assessment (HRA), represented by methods such as THERP (Technique for Human Error Rate Prediction) and HCR (Human Cognitive Reliability), all assumed that it was meaningful to refer to a human error probability, although they also acknowledged that the value or magnitude of this depended on external performance shaping factors. The human failure probability was nevertheless the coveted 'signal', while the influence of the performance shaping factors was the 'noise'. This position was later effectively reversed in the so-called second generation HRA methods. In these methods, represented by ATHEANA (A Technique for Human Error Analysis), CREAM (Cognitive Reliability and Error Analysis Method), and MERMOS (Méthode d'Evaluation de la Realisations des Missions Opérateur pour la Sûreté), the influence of the performance conditions was seen as more important than the postulated human error probability. In other words, the influence of the performance shaping factors now became the signal, while the human error probability became the noise, to the extent that some methods even refrained from referring to the notion of human error at all.

A similar transition took place when the focus changed from the human factor to the organisation and/or safety culture. This happened in the mid 1980s, and is often linked to the disaster at the Chernobyl nuclear power plant and the loss of the space shuttle *Challenger* (both in 1986). In order to understand these accidents it became necessary to introduce new factors or conditions, although the basic thinking about safety remained the same. But even though the idea of safety culture was useful, it was nevertheless difficult to include the organisation in the calculation of failure probabilities. In practice, the search for a 'human error probability' was complemented by a search for an 'organisational error probability' or 'organisational failure rate' – although it was not usually expressed so bluntly. Describing it in this way, however, makes it clear that an 'organisational failure rate' is a meaningless concept. An organisation can neither fail nor function in the same way that a component can; that is, it does not make sense to think about it in terms of the bimodal principle, as being either right or wrong. Indeed, an organisation – or a department in an organisation or a specific role – cannot really be thought of as a component in the first place.

Although HRA nominally looked for the probability of a 'human error', the focus was actually on how human performance of a specific function might fail to reach its objectives rather than whether the human as such failed. In practice, the terminology nevertheless (mis) led people to focus on error probabilities. While it may be justified in the technological domain to see the performance of a function as synonymous with the state of a component, it is clearly not so in the case of human functions or the functions of a socio-technical system.

The differences in perspective become clear when a system is defined in terms of how it functions rather than in terms of its architecture and components. In this case the question is whether the functioning achieves its purposes. But this cannot be simplified to a question of whether the system is in a 'normal' state or a 'failed' state. It is instead a question of the variability of functioning and whether the outcome is acceptable under the existing conditions. But as soon as we say variability, we also acknowledge that any 'failure' will be temporary, hence reversible. We should consequentially try to understand how likely the variability of a system's performance is, and how the variability of multiple functions may interact to produce an unintended – and in most cases also unwanted – outcome.

What we are interested in is, however, not whether a function will be variable, since by definition they all are. Instead we are interested in whether the variability will be so large that the function will be unable to provide the desired outcomes. This can be due either to the variability of a single function or – more likely and also more importantly – the combination of the variability of multiple functions over time and over space. There will of course always be cases (even in complex socio-technical systems) where the variability of a single function (or a single activity) is so large that an adverse outcome is inevitable. But even in that case it is of limited use to say that the function has failed or that the component or entity has malfunctioned and to calculate the probability that this happened or will happen. In most cases the (unwanted) outcomes are due to interactions among individual functions, hence combinations of the effects of their variability. That being the case, it is clearly necessary to find ways to identify the potential for variability and to analyse how this may combine to produce the strong signals or the unwanted effects.

## Conclusions

Virtually all accident investigations and risk assessments are conducted in a state of relative ignorance of the full behaviour of the system. This condition contrasts with the fact that all established approaches to risk assessment require that it is possible to describe the system and the scenarios in detail; that is, that the system is tractable. Unfortunately, all socio-technical systems are more or less intractable, which means that the established methods are not suitable. Since it is not reasonable to overcome this problem by making system descriptions so simple that they become tractable, it is necessary to look for approaches that can be used for intractable systems, that is, for systems that are incompletely described or underspecified.

Resilience Engineering provides the basis for such approaches. Resilience Engineering starts from a description of characteristic functions, and looks for ways to enhance a system's ability to respond, monitor, learn and anticipate. By emphasising that safety is something a system does rather than something it has, the unavoidable state of relative ignorance can be reduced by focusing on what actually happens. To do so requires a set of concepts, a terminology and a set of methods that make it possible to describe work-as-done rather than work-as-imagined. That is what the FRAM is about.

## Comments on Chapter 1

The quote at the beginning is from page 93 in Petroski, H. (1994), *Design Paradigms*, Cambridge University Press.

The first presentation of the What-You-Look-For-Is-What-You-Find (WYLFIWYF) principle was at a Resilience Engineering workshop in Rio de Janeiro in December 2005. An explanation and examples can be found in Lundberg, J., Rollenhagen, C. and Hollnagel, E. (2009), 'What-You-Look-For-Is-What-You-Find – The consequences of underlying accident models in eight accident investigation manuals', *Safety Science*, 47, 1297–1311.

The full Coleridge quotation from 1835 is: *'If men could learn from history, what lessons might it teach us! But passion and party blind our eyes, and the light which experience gives us is a lantern on the stern which shines only on the waves behind us.'* Søren Kierkegaard wrote that *'Livet forstås baglænds, men må leves forlænds'* in 1843.

The lack of imagination in designing, analysing and managing both technological and socio-technical systems can easily have disastrous consequences and the importance of sufficient – or requisite – imagination can hardly be overstated. An early argument for that can be found in Westrum, R. (1991), *Technologies and Society: The Shaping of People and Things*, Belmont, CA: Wadsworth.

Complexity, particularly as in complex systems, is often invoked as a *deus ex machina* to 'explain' why the use of modern technology is marked by so many unexpected events. Many also seem captivated by the so-called Complexity Sciences, although this may turn out to be the triumph of hope over experience. A good discussion of complexity is provided by Pringle, J. W. S. (1951), 'On the parallel between learning and evolution', *Behaviour*, 3, 175–215.

There are several descriptions or surveys of HRA models, which also include part of their history. For example Kirwan, B. (1994), *A Guide to Practical Human Reliability Assessment*, London: Taylor & Francis, or Hollnagel, E. (1998), *Cognitive Reliability and Error Analysis Method (CREAM)*, Oxford: Elsevier Science Ltd.

# Chapter 2
# The Intellectual Background

The real trouble with this world of ours is not that it is an unreasonable world, nor even that it is a reasonable one. The commonest kind of trouble is that it is nearly reasonable, but not quite. Life is not an illogicality; yet it is a trap for logicians. It looks just a little more mathematical and regular than it is; its exactitude is obvious, but its inexactitude is hidden; its wildness lies in wait...

(Chesterton, 1909)

## The Naturalness of Linear Thinking

It is natural and nearly irresistible to think of events as if they develop in a step-by-step progression, where one action or event follows another. In technical terms this is called linear thinking. Events do, of course, happen one by one in the sense that two events cannot happen at the same time, at least if time is measured precisely enough. (I am, of course, referring to events that belong together in the sense of being part of the same development or situation, rather than events that are completely independent of each other.) Two events may happen in the same year, in the same month, on the same day, and even at the same time of day. But by increasing the temporal resolution, it is always possible to place one event earlier (or later) than the other. A good illustration of that is how sporting events, such as the 100 metres dash, are timed. In the 1920s it was sufficient to time runners to a tenth of a second. Today the timing is done to a thousandth of a second, in order to determine whether one athlete completes the distance before another.

Linear thinking is, however, not only the idea that events happen one after another, but also that there is a cause–effect relationship between them. It is, of course, intuitively sensible that one thing can lead to another. Indeed, it is the very basis for our ability to deal with the world around us, to build complicated machines, and to operate large-scale industrial installations. We know that when A happens, that I turn the ignition key in my car, then B will also happen, that is, the engine will start. The same way of reasoning is also applied in the opposite direction, meaning that when something happens (an effect), then we assume that something else must have happened shortly before (a cause).

We can find this thinking at the very roots of Western culture. Leucippus of Miletus, c.480–c.420 BC, said that 'Nothing happens in vain, but everything from reason and of necessity.' At about the same time, Aristotle, 384–322 BC, wrote that 'all causes of things are beginnings; that we have scientific knowledge when we know the cause; that to know a thing's existence is to know the reason why it is'.

Linear thinking is not only used when we design and build systems, but also when the unexpected happens. When something goes wrong we assume that we can find the reason for it: something must have failed or malfunctioned to bring about the unwanted outcome. It is this way of thinking that makes us focus on components and component failures or malfunctions, as in the Domino model, in Root Cause Analysis, and Human Reliability Assessment.

This simple way of thinking was not unreasonable when technological systems were relatively easy to understand in the nineteenth century. In some sense, there was little or no ignorance when systems (machines) were built and constructed – at least about how the machines worked (there may have been hidden risks related to the materials, as the history of materials and chemistry bears witness to). Even when the unexpected happened because materials failed, the reason was understandable, that is, it was a simple cause, a failure of something, and not an intricate web of causes. Technological systems are, however, no longer simple and easy to understand. Today, when we furthermore must deal with socio-technical rather than technical systems, the simple linear way of thinking is no longer reasonable.

## Simple Linear Thinking

Humans have tried to understand the world around them from the beginning of time, and certainly from the beginning of written history. As psychologists and philosophers never tire to explain, we have a need of certainty, of feeling in control. The German philosopher Friedrich Wilhelm Nietzsche (1844–1900), expressed it as follows:

> (t)o trace something unknown back to something known is alleviating, soothing, gratifying and gives moreover a feeling of power. Danger, disquiet,

anxiety attend the unknown – the first instinct is to eliminate these distressing states. First principle: any explanation is better than none...

In order to accomplish this, the models, the classifications, the principles, the concepts that we use to explain what goes on around us must be in accord with reality. If the world around us was constant or stable, then the understanding would gradually become fine-tuned to the context, and we would be able perfectly to understand everything – or at least not suffer from any serious ignorance. But the world around us is not constant. It has always been changing, at first so slowly that it did not pose a problem, but since the 1950s so rapidly and with ever increasing speed that we no longer are able to keep pace.

One reason for this is that humans have become overconfident in their ability to build larger and larger systems. A contributing factor has no doubt been the 'logic' of economy of scale, that is, the dictum that it is cost-effective to build larger machines and larger systems. Human endeavour has always nourished a demand for faster, better and cheaper systems even decades before the NASA achieved a dubious reputation for espousing this principle. The pressure to build systems that were faster, better and cheaper, combined with the lack of imagination that is a result of habituation and (organisational) amnesia, meant that humans at some point between the 1950s and 1980s passed a threshold where they no longer knew exactly what they were doing. But having made themselves dependent on such systems, there was no way of stopping the development.

## Complex Linear Thinking

The shortcomings of simple linear thinking became commonly acknowledged in the late 1970s. It was found that serious events could happen even in relatively well-controlled work environments and that such events often involved multiple sequences occurring in series and in parallel. Organisations that were able successfully to handle the simpler situations (the so-called single failure criterion), would still remain vulnerable to multiple simultaneous failures. In order to cope with these, accident investigation, risk assessment and safety management had to go beyond simple linear thinking and beyond simplified categories of conditions and acts. As a matter of fact, the need for this kind of change was argued much earlier. In 1956, Morris Schulzinger summarised a 20-year study of 35,000 accidents by noting that accidents resulted from 'the integration of a dynamic variable constellation of forces and occurs as a sudden, unplanned and uncontrolled event'.

This, and similar writings, pointed out that the sequences were complex and interacting, decades before the famous Swiss cheese model did the same. A proper understanding of safety could no longer rely on simple sequences of causes and effects as a way to explain past and future risks, but required a synthesis of technological, psychological, organisational, environmental and temporal factors. This did eventually lead to models and methods comprising multiple sequences and latent conditions, although the method development lagged far behind the practical needs. And the basis for the models was still the linear thinking that implied both order and causality.

## Dynamic Systems and Parallel Developments

There are probably good psychological and philosophical reasons why humans seem to prefer simple linear thinking when they try to understand why things have gone wrong – as in accident and incident investigations. The same is the case for thinking about what may happen in the future, as in risk assessment, where the future paradigmatically is represented as a sequence of events where each event can either succeed or fail; the standard event tree is an example of that.

It is, however, important to separate the fact that events always can be ordered in time from the assumption of causality. That event B succeeds event A only means that they are ordered in time, not that A is the cause of B, as noted by the Scottish philosopher David Hume more than 260 years ago. The issue of how to represent events in order to understand how and why they happen has been the subject of much thinking since then, not least by psychologists. Psychology studies behaviour, that is, what people do individually and collectively, and tries to find explanations for what is observed. The dominating school of thought was for many years that humans, and other living organisms, responded to given conditions or situations (called stimuli) in ways that could be described precisely or even mathematically. In other words, behaviour – including acting, thinking and feeling – could be described and explained in terms of chains of stimuli (causes) and responses (effects). This behaviourist school of thinking came about at the beginning of the twentieth century, and dominated mainstream psychology for at least five decades. It established the tradition of quantifying behaviour, something that agreed well with the later engineering approach to safety.

The assumption that behaviour and performance can be described as a series of individual events, in which each succeeding effect becomes a cause for the next effect, nevertheless raises a serious problem. When we study a specific event, or a short series of events, then how should we delimit the number of cause–effect pairs that precede the series or that follow it? Indeed, there is no logical reason why the causal series should not be extended indefinitely at both ends.

A pragmatic solution to this problem, presented by the social psychologist Floyd Allport in 1954, was to consider a 'certain' part of the chain as belonging to a particular behavioural act and to ignore the rest. That this could be done without invoking accusations of any arbitrariness was seen by Allport as evidence that some other principle than linear causation was at work. He specifically made the observation that:

> Patterns seem to flow not from linear trains of causes and effects, but somehow from patterns already existing. Structures come from structures; and in many cases the structures themselves, as wholes, seem to operate not sequentially but in a contemporaneous or concurrent fashion.

The solution was to replace the idea of sequences of events with the idea of patterns of events, where each pattern closed itself through a cycle of operation. Allport concretely described an approach to analysis and representation based on event-structures, and called it event-structure diagrams. A crucial part of this was the notion of tangent cycles or tangent structures, that is, event-structures that could exist and affect other event-structures by augmenting or decreasing their energies, rather than by cause–effect relations. (The idea of energy, energies or even energetic units was in line with the scientific thinking of the 1950s, not least the emerging science of cybernetics, of which more will be said in the following.)

The event-structures provided a way of explaining how things happened and how different events could relate to each other that avoided the issue of a potentially infinite regress to indefinite origins. It also avoided the boundary problem, since event-structures were defined as patterns of *events* rather than as set or subsets of *components*. As part of the intellectual background for the FRAM, the event-structure theory demonstrated that it was possible to account for behaviour without invoking cause–effect relations. Allport's notion of cycles as the constituent of events is in principle the same as the functions in the FRAM. The notion of tangential (tangent) cycles is similar to the principle of resonance described in Chapter 3, since both are of a non-causal nature. And finally, abandoning the hierarchical structures and the fixed ordering of events for an emphasis on equilibrium or 'static' structural conditions is similar to the notion of potential couplings in a FRAM model.


## The Second Cybernetics

The science of cybernetics was introduced by Norbert Wiener in 1948 – although many people today use the 'cyber' prefix in blissful ignorance of the origin. Cybernetics was developed to study control and regulation, and one of the fundamental concepts was the feedback loop. A feedback loop describes how a measurement of the output from the system can be fed back and used as a basis for regulating the process that produces the output. The classical example is how a thermostat regulates the heating (or cooling) of a room. If the measured temperature is too low, the thermostat switches on the heating unit; if the measured temperature is too high, the heating unit is switched off. The thermostat thus makes the system self-regulating by counteracting deviations in order to ensure that the output stays within a specified range.

A slightly more complex example is the way in which central banks, such as the European Central Bank, try to regulate the economy of a country or region, see Figure 2.1. If the economy 'overheats', as measured by the inflation rate, central banks typically respond by increasing the interest rate because this is assumed to slow down the economy. Conversely, if the economy 'cools,' the interest rate is reduced, as this is assumed to stimulate the economy. The system, meaning the national economy and the central bank, is self-regulating and deviation-counteracting – at least to the extent that the underlying assumptions are correct.

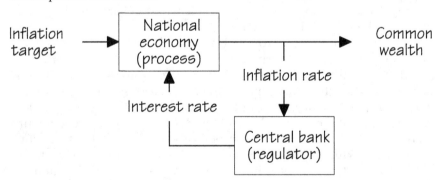

**Figure 2.1      Control of the economy by negative feedback**

While the negative feedback loop effectively can counteract deviations in the underlying process, at least within certain limits, there are also situations where the regulator responds so that the original signal is amplified instead of dampened. This was pointed out by the mathematician Margoroh Maruyama, who in 1963 argued for the need of what he called the 'Second cybernetics'.

> By focusing on the deviation-counteracting aspect of the mutual causal relationships however, the cyberneticians paid less attention to the systems in which the mutual causal effects are deviation-amplifying. Such systems are ubiquitous: accumulation of capital in industry, evolution of living organisms, the rise of cultures of various types, interpersonal processes which produce mental illness, international conflicts, and the processes that are loosely termed as 'vicious circles' and 'compound interests'; in short, all processes of mutual causal relationships that amplify an insignificant or accidental initial kick, build up deviation and diverge from the initial condition.

This is highly similar to the ideas about how event-structures can affect each other by augmenting or decreasing their 'energies', as described above. The use of cybernetics, and especially the 'second cybernetics', makes it possible to represent the event-structures as closed loops and thereby to be more precise about how the 'mutual causal relationships' work. The world of finance offers a good example of that. Figure 2.2 shows two coupled loops. The one on the left describes the couplings between 'relative value', 'demand', and 'price', while the one on the right describes the couplings between 'price', 'profits', and 'supply'. The symbol ⊕ (and full black lines)

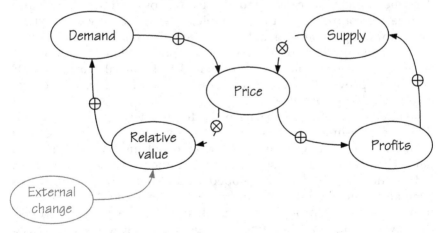

**Figure 2.2**     **Mutually coupled processes**

means that the relation between two entities is proportional while the symbol ⊗ (and dashed black lines) means that the relation is inverse.

Both main loops in Figure 2.2 are self-regulating. For instance, if 'relative value' increases, then 'demand' also increases. If 'demand' increases then 'price' increases, following which the 'relative value' decreases. And if 'relative value' decreases, then 'demand' also decreases and so on. For the other loop, if 'profits' increase, then 'supply' also increases. If 'supply' increases then 'price' decreases, following which the 'profits' decrease. And if 'profits' decrease, then 'supply' also decreases and so on.

While either of the two main loops is self-regulating in the sense that each would reach a stable equilibrium after some time, the two loops put together does not show the same kind of stability. (Hint: the left-hand loop affects 'price' in the opposite direction of the right-hand loop.) The relationships that so aptly are represented in Figure 2.2 would be very difficult to capture with linear sequences, whether simple or complex.

## Conclusions

This chapter has argued that neither simple linear thinking, nor complex linear thinking, suffice to describe the world around us. This was realised by scientists more than 50 years ago but had few, if any, consequences for how safety was practised. It was not until the 1990s that people were forced to recognise that they no longer were dealing with tractable, homogeneous worlds that could be understood and described in terms of causes and effects. At about that time they found themselves confronted with intractable, heterogeneous worlds where linear thinking came to naught. The second cybernetics had shown analytically how this could lead to situations where control was lost and outcomes were out of proportion. The financial world in 2008 obligingly provided a convincing example of what a loss of control could look like in practice.

While the intellectual background for what we now would call dynamic, non-linear models was in place by the mid 1960s, the need to use such models were not fully recognised until several decades later. Historically, the classical safety methods, such as Fault Trees, FMEA and HAZOP, were developed around the same period. It was generally believed that if the technological risks could be accounted for, for example, through a Probabilistic Safety Assessment (PSA), then everything would be fine. That belief was severely challenged by the accident at Three Mile Island (TMI) nuclear power plant in 1979. After TMI the need to include the human factor soon became the accepted wisdom. This quickly led to a proliferation of 'new' methods that nevertheless were based on the same principles – decomposition and cause–effect links. The occurrence of the so-called organisational accidents, Challenger and Chernobyl both in 1986, further emphasised the necessity of being able to account for the non-technical parts of the systems (which now were called socio-technical systems). But the thinking remained on the same track and it was not until the end of the twentieth century that the need for an alternative approach slowly became accepted, leading to the proposal for Resilience Engineering. As part of that development, the FRAM shows how it is possible to do safety analyses without decomposing systems into components and without being dependent on the notion of causality.

## Comments on Chapter 2

The quote at the beginning is from Chesterton G. K. (1909), *Orthodoxy*, New York: Lane Press. The Nietzsche quote a few pages later can be found on page 93 in Nietzsche, F. W. (2007; orig. 1895), *Twilight of the Idols*, Ware, Hertfordshire: Wordsworth Editions Limited.

Morris S. Schulzinger's study was published in a book entitled '*The Accident Syndrome: The Genesis of Accidental Injury*', Illinois: Charles C. Thomas Publishers. The data for the study was collected during 1930–1948, and described about 35,000 accidental injuries.

Floyd Allport (1890–1978) is not well known by the general public, but is often considered the founder of social psychology as a scientific discipline. His discussion of the events is in a paper from 1954 entitled 'The structure of events: Outline of a general theory with applications to psychology', *Psychological Review*, 61(5), 281–303. What Allport called 'structure' and 'event-structure' correspond to what in today's terminology would be called functions and homeostatic loops.

Norbert Wiener introduced the new science of cybernetics in 1948 by a book entitled *Cybernetics, or Control and Communication in the Animal and Machine*. While the importance of cybernetics for science – and literature – can hardly be overstated, the so-called Second Cybernetics (not to be confused with second-order cybernetics) remains relatively unknown. It was proposed by M. Maruyama in 1963, in a paper entitled 'The second cybernetics: Deviation-amplifying mutual causal processes,' *American Scientist*, 55, 164–179. There are significant similarities between the second cybernetics and the System Dynamics approach to the behaviour of complex systems, developed by Jay Forrester (1918–).

# Chapter 3
# The Principles

Though this be madness, yet there is method in't.

(Hamlet, Act II, Scene 2)

The FRAM has been developed for a specific purpose more or less from scratch. The purpose was to provide a method that recognised successes as the flip side of failures – in other words, a method that focused on the nature of everyday activities rather than on the nature of failures. The method should also be able to analyse past events as well as possible future events – specifically, but not exclusively, what might go wrong. The reason for this is simply that things happen in the same way regardless of whether they are actual past events or possible future events. We must, by necessity, always look at what happens from an arbitrary point in time that we call the present or 'now'. Relative to that, some events have happened (the 'past') and some events are yet to happen (the 'future'). But it would be strange, indeed, if the way in which something happens changes the moment it passes through the 'now', that is, when a possible future event becomes an actual past event. Yet that is precisely what seems to be the case, at least as far as current safety approaches are concerned, since models and methods for accident investigation differ from models and methods for risk assessment.

In the development of the FRAM nothing was taken for granted. As described in Chapter 10, the FRAM is also a *method-sine-model* rather than a *model-cum-method*, that is, its purpose is to build a model of how things happen rather than to interpret what happens in the terms of a model. There is obviously an intellectual background, the most important of which has been described in Chapter 2, as well as the even more fundamental tradition of critical thinking that has dominated science for more than 2,500 years. But as there were no comparable methods for safety assessment, even taking the meaning of safety quite broadly, formulating the underlying principles from scratch made a virtue out of necessity.

The four principles on which the FRAM is built are:

- That failures and successes are equivalent in the sense that they have the same origin. Another way of saying that is that things go right and go wrong for the same reasons.
- That the everyday performance of socio-technical systems, including humans individually and collectively, always is adjusted to match the conditions.
- That many of the outcomes we notice – as well as many that we do not – must be described as emergent rather than resultant.
- Finally, that the relations and dependencies among the functions of a system must be described as they develop in a specific situation rather than as predetermined cause–effect links. This is done by using functional resonance.

Each of the four principles will be explained in detail in the sections that follow.

## The Equivalence of Failures and Successes

The starting point for thinking about safety is usually that something has gone wrong, which means that something has happened that was not supposed to happen. The starting point is thus an unexpected outcome possibly preceded by an unwanted event, and the safety quest serves to find the reasons for that.

Since we are more concerned about the unexpected and unknown than the habitual and known, it is not surprising that we spend considerable efforts to understand the former and almost none to understand the latter. This has the unfortunate consequence that we know relatively little about how things go right, and that we often simply assume that things go right because systems are well designed, well tested and well behaved. The threats to normal performance are therefore seen as coming from various types of failures or malfunctions, equipment faults and in particular human errors and violations. Most efforts are accordingly directed at understanding why things go wrong.

However, whenever something is done, the intention is always to do something right and never to do something wrong. (Even if a person wants to do something harmful, it is still necessary to do it 'right'.) Assume, for the sake of discussion, that performance can be described as a sequence of actions. For each action, the choice of what to do is determined by many different things, including competence, understanding of the situation, experience, habit, demands, available resources and expectations about how the situation may develop – not least about what others may do. If the expected outcome is obtained, the next action is taken and so on. But if the outcome is unexpected, then the preceding action is re-evaluated and classified as wrong rather than right, as an error or as a mistake, using the common but fallacious *post hoc ergo propter hoc* argument. With hindsight, it is pointed out what should have been done, if only people had made the necessary effort at the time. The whole situation is neatly summarised in Figure 3.1. The whole argument is, however, unreasonable because the action was chosen based on the expected rather than the actual outcome. It was therefore not the action *per se* that was wrong, but rather the action – or the expectations – relative to the situation at the time.

The unreasonableness of this argumentation was succinctly pointed out by the physicist Ernst Mach, who in 1905 wrote that 'knowledge and error flow from the same mental sources, only success can tell one from the other'. Failures and successes are thus equivalent in the sense that we can only say whether the preceding action was right or wrong after the outcome is known. Despite the obvious reasonableness of this argument, safety practices have stubbornly focused only on that which goes wrong.

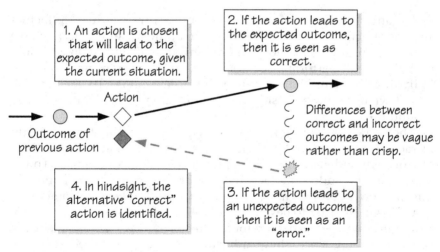

**Figure 3.1     Re-evaluation of actions with unexpected outcomes**

## The Approximate Adjustments

Technological artefacts, large and small, are to the best of our abilities designed, built and maintained to produce a near constant performance

– at least until something fails. The same is, unfortunately, not the case for humans or for social systems (organisations). Human performance is always variable, due to a number of internal and external factors, such as:

- Inherent physiological and/or psychological characteristics. Examples are fatigue, circadian rhythm, vigilance and attention, refractory periods, forgetting and so on.
- Higher level psychological phenomena such as ingenuity, creativity and adaptability used, for instance, to overcome temporal constraints, under-specification and boredom.
- Organisational factors, such as external demands (to quality or quantity), deadlines leading to resource stretching, goal substitution, organisational double-binds (for instance safety-productivity conflicts) and so on.
- Social factors, such as the expectations to oneself or to colleagues, compliance with group working standards and so on.
- Contextual factors, such as working conditions that can be too hot, too noisy, too humid, too crowded and so on.
- Unexpected changes in the work environment; for example, weather, technical problems and so on.

Organisational performance variability can be due to factors such as effectiveness of communication, authority gradient, (lack of) trust, organisational memory and organisational culture. In addition, organisational performance variability may also come from the individual humans that populate the organisation. To that can be added what happens outside the organisation, that is, in the world around it.

The work situation in large-scale socio-technical systems is partly intractable, as described in Chapter 1. This means that the conditions of work are underspecified, in principle as well as in practice. To the latter must be added the fact that resources – materials, manpower, information and especially time – usually are limited and sometimes insufficient. In order to be able to do their work, people – individually and collectively – must therefore adjust what they do to match the conditions. But since these adjustments are needed because the situation is underspecified and because resources are insufficient, the adjustments will themselves be approximate rather than precise. They are nevertheless usually good enough to get the job done, and are in fact the reason why things go right far more often than they go wrong, that is, the reason why everyday performance is successful. Yet approximate adjustments are also the reason why things something go wrong. This is an additional argument for the Equivalence principle described above.

Performance variability is on the whole a strength rather than a liability, and is often the primary reason why socio-technical systems function as well as they do. Humans are extremely adept at finding effective ways of overcoming problems at work, and this capability is crucial for both safety and productivity. Since human performance can both enhance and detract from system safety (and quality) assessment methods must be able to address this duality.

## Emergence

Whenever something happens, in a system or in the world at large, whenever there is an unexpected event, an explanation is sought. In most cases the explanation relies on the general understanding of how systems function, which means that it includes the principles of decomposition and causality. In such cases the outcome is said to be a result of the 'inner' workings of the system and is therefore technically called *resultant*.

There are, however, a growing number of cases where it is not possible to explain what happens as a result of known processes or developments. In such cases the outcome is said to be *emergent* rather than *resultant*. It is still possible to provide an explanation of what happened, but the explanation will be of a different nature. The meaning of emergence is not that something happens 'magically', but simply that it happens in such a way that it cannot be explained using the principles of decomposition and causality. This is typically the case for systems that in part or in whole are intractable.

The use of the term 'emergent' acknowledges that an explanation in terms of causality is inappropriate – and perhaps even impossible. The first known use of the term was by the English philosopher George Henry Lewes (1817–1878), who described emergent effects as not being additive and neither predictable from knowledge of their components nor decomposable into those components. In the contemporary vocabulary this means that the effects are non-linear and that the underlying system is partly intractable.

### Elusive Causes

Whatever happens is, of course, real in the sense that there is an effect or an outcome, an observable change of something – sometimes described as a transition or a state change. It may be that the battery in a laptop is flat, that a report has been delivered, that a pipe is leaking or that a problem has been solved. The outcome may also be a sponge forgotten in the abdomen of a patient, a capsized vessel or a financial meltdown. Classical safety thinking assumes that the causes are as real as the effects – which makes sense according to the logic that the causes in turn are the effects of other causes one step removed and so on. The purpose of accident and incident investigation is therefore to trace the developments backwards from the observable outcome to the efficient cause. Similarly, risk assessment projects the developments forward from the efficient cause to the possible outcomes.

A simple example is the investigation of a traffic accident, for instance that a car has veered off the road and hit a tree. An investigation may determine that the accident was due to a combination of the following: that the road was not well-maintained, that the driver was drunk, that the tyres were worn out and that the weather was wet and stormy. Furthermore, these conditions not only existed when the accident happened, but may also to some extent exist afterwards. The (wrecked) car can be inspected and the condition of the tyres established. The same goes for the road surface. We can find out how the weather was by consulting with the meteorological office. And in many cases the level of drunkenness of the driver may be determined from physiological tests. In other words, both the effects and the causes are 'real', and the outcome can therefore rightly be said to be a result of these causes. The reality of the causes also makes it possible to trace them further back and possibly do something about them, eliminating them, isolating them, protecting against them (or rather their effects) and so on.

In the case of emergent outcomes, the causes are, however, elusive rather than real. The principle of emergence means that while the observed (final) outcomes usually are permanent or leave permanent traces, the same is not the case for what brought them about. The outcomes may be due to transient phenomena or conditions that only were present at a particular point in time and space, as combinations that existed for one brief moment. These conditions may, in turn, be explained by other transient phenomena and so on. Causes therefore have to be reconstructed rather than found. This means that we may not be able to do something about them in the usual manner, but we may possibly be able to control the conditions that brought them into existence, provided these are sufficiently regular and recurrent. We can, of course, also protect against these conditions by devising various forms of prevention and protection.

The principle of emergence means that while we may plausibly argue that a condition existed at some point in time, we can never be absolutely sure. However, the variability of performance is 'permanent' in the sense that there will always be some variability. The variability is furthermore systematic rather than random, hence largely predictable. This is why we can base safety analyses on the existence or presence of variability.

## Resonance

The fourth and final principle of the FRAM is that the couplings or dependencies among the functions of a system must be accounted for as they develop in a specific situation. One implication is that the couplings cannot be specified precisely in advance, hence not be represented by a tree or network structure. A second implication is that the dependencies can go beyond simple cause–effect relations and be more like the event-structures described in Chapter 2. Indeed, the third principle – Emergence – means that there will be outcomes that cannot be understood in terms of causality, and that a more powerful principle of explanation therefore is required. This is why the fourth principle uses functional resonance to explain what can happen in complex socio-technical systems. To understand that, it is necessary to begin by a short description of classical resonance.

### What is Resonance?

In physical systems, classical (or mechanical) resonance refers to the phenomenon that a system can oscillate with larger amplitude at some frequencies than at others. These are known as the system's resonant (or resonance) frequencies. At these frequencies even small external forces that are applied repeatedly can produce large amplitude oscillations, which may seriously damage or even destroy the system. The most famous examples are bridges that have collapsed, such as the Tacoma Narrows bridge in 1940.

A simple example of resonance is a child on a swing. Here the movement (amplitude) of the swing can increase either if a kind adult provides a regular push to the child/swing, or if the child has learned how to increase the motion of the swing. A child on a swing nicely illustrates classical resonance, and how the effect develops over time. In the case of intended amplification, the signal or oscillation can be very small but must nevertheless be above the threshold of detection, since it would otherwise be impossible to synchronise the external forcing function to it.

Whereas classical resonance has been known and used for several thousand years, stochastic resonance is of a more recent origin. In stochastic resonance there is no forcing function, but rather random noise, which every now and then pushes a subliminal signal over the detection threshold. Stochastic resonance can be defined as the enhanced sensitivity of a device to a weak signal that occurs when random noise is added to the mix. The outcome of stochastic resonance is non-linear, which simply means that the output is not directly proportional to the input. The outcome can also occur – or emerge – instantaneously, unlike classical resonance which must be built-up over time.

The concept of stochastic resonance can be used to understand how unexpected outcomes happen. It was originally applied to climate change, and has also been used to describe events in the financial markets. On a much smaller scale it has been used in neurobiology to describe what happens in the sensory system. It temptingly offers a hope of understanding some of the large-scale unexpected outcomes in safety critical systems that seem to defy conventional explanations. In that respect the weakness of stochastic resonance is, however, that it is stochastic or random. For the purpose of improving safety, it is necessary to be more precise and preferably to be able to predict what may happen in a deterministic rather than a probabilistic sense.

## From Stochastic Resonance to Functional Resonance

The second principle of the FRAM was the principle of performance variability. The fact that human and organisational performance always is adjusted, and adjusted approximately, to match the conditions of work is another way of saying that there always is performance variability. This performance variability is in most cases so small that it does not have any unwanted consequences – and so small that it hardly is noticed. Resilience engineering proposes that this is the reason why things go right, hence usually a strength rather than a liability. The FRAM uses stochastic resonance as an analogy to propose how everyday performance variability can lead to unexpected outcomes. If we focus on the performance of any specific part of a system, a function carried out by an individual, by a joint cognitive system, by a group or by the organisation, the performance variability of that function will usually be subliminal, hence not noticeable. Relative to this subliminal signal, the performance variability of the rest of the system, that is, the combined performance variability of all the other parts, can be seen as random noise. In stochastic resonance terms, the – random – performance variability in the rest of the system may interact with the subliminal signal and make it superliminal or detectable. This offers a way to understand how noticeable outcomes apparently can arise or emerge out of nothing, without invoking cause–effect relations.

The FRAM, however, refers to functional resonance rather than stochastic resonance. The reason for this is simply that the variability of performance in a socio-technical system is not as random as it may seem to be. The variability is in the main due to the approximate adjustments of people, individually and collectively, and of organisations that are the basis of everyday functioning. These approximate adjustments are not only purposive, but also make use of a small number of recognisable short-cuts or heuristics. Performance variability, in other words, is semi-regular or semi-orderly and therefore also partly predictable. There is a regularity in how people behave and in the way they respond to unexpected situations – including those that arise from how other people behave.

The latter is particularly important and constitutes the main difference between stochastic and functional resonance. Performance variability is not merely reactive but also – and more importantly – proactive. People not only respond to what others do but also to what they expect that others will do. People often respond ahead of time in ways that are systematic and partly predictable. The approximate adjustments themselves are thus made in response – and anticipation – of what others may do, individually or collectively. The predictability may not be perfect, but the variability is definitely not random. What each person does obviously becomes part of the situation that others adjust to – reactively and proactively. This gives rice to a dependency and means that the approximate adjustments become *mutual approximate adjustments*. In other words, the functions in a system become coupled, and the performance variability of the functions therefore also becomes coupled.

Functional resonance can more formally be defined as the detectable signal that emerges from the unintended interaction of the everyday variability of multiple signals. The signals are usually subliminal, both the 'target' signal and the combination of the remaining signals that constitutes the noise. But they are all subject to certain regularities that are characteristic for different types of functions, hence not random or stochastic. The resonance effects that occur can be seen as a consequence of the ways in which the system functions, and the phenomenon is therefore called functional resonance rather than stochastic resonance. Functional resonance is proposed as a way to understand outcomes that are both non-causal (emergent) and non-linear in a way that makes both predictability and control possible.

## Frequency and Variability

In the FRAM, variability can be considered as noise. In the theory of stochastic resonance – turned into the principle of functional resonance – the variability is the source of the random noise that is the basic element of the resonance phenomenon. But the distinction between signal and noise is relative to the duration of the event being considered (the sampling period). While safety traditionally has focused on scenarios where the performance of human variability was significant, hence constituted a signal, this need not be so. Indeed, it is important to acknowledge that while this conventional focus is understandable, it is not compulsory.

There is, of course, a good reason for focusing on human performance, namely that the practical interest for safety has been related to situations of work (industry), where human performance was or is a crucial constituent. The focus was formalised in the beginning of the twentieth century and was reinforced by momentous events such as the disaster at the Three Mile Island nuclear power plant in 1979. This focus was sustained until the mid 1980s, when several major disasters, such as the explosion of the space shuttle *Challenger* and the accident at the Chernobyl nuclear power plant, forced people to acknowledge that other factors also played a role. This led to a change in focus from human to organisational performance. Initially it meant that the organisational influences of human performance were taken into account, and were acknowledged as important, 'new' influences or factors. But in the longer term the focus changed from the performance of people (whether individually or collectively) to the performance of organisations.

When the focus is on human activity, that which happens in the organisation – in the background or at the blunt end – may take place so slowly or change so slowly that it for all intents and purposes can be regarded as constant relative to the focus. This basically means that what happens in the background can be treated as something constant, as a performance shaping factor, hence more or less disregarded by the analysis as such. However, if the focus is on the performance of the organisation, the opposite is the case. In other words, when studying phenomena such as organisational changes (or even more long-term developments, such as cultural, societal or environmental changes), the variability of human performance can be considered as high-frequency background noise. So even though the variability of human performance may be significant for the study of short-term events, such as an 8-hour shift, we may safely disregard it for studies that look at much longer durations.

## Conclusions

This chapter has presented the four principles that constitute the conceptual foundation of the FRAM. They are the principle of the equivalence of failures and successes, the principle of approximate adjustments, the principle of emergence and the principle of functional resonance. The four principles can be seen as a summary of the general experiences from safety studies since the late 1980s, not least the increasing dissatisfaction with the ability of established methods to deal with the complexity of socio-technical systems, a conditions that is only likely to become worse.

One realisation is that safety models and methods must go beyond simple cause–effect relations. There are a growing number of accidents that can be better understood as resulting from an alignment of conditions and occurrences than as direct consequences of failures and malfunctions. Another realisation is that the causes of accidents and incidents are constructed rather than found. It is therefore more important to understand the nature of system dynamics and the variability of performance, than to model individual technological or human failures. Human actions, in particular, cannot and should not be treated in isolation.

The experiences from studying work-as-done – rather than work-as-imagined – is that all social systems, from individuals to organisations and nations, can be seen as trying to balance efficiency and thoroughness both on the level of specific actions and on the level of overall strategies. One reason for these trade-offs or adjustments is the need to absorb the effects of the everyday performance adjustments that are made by everyone else in the system.

## Comments on Chapter 3

The FRAM uses resonance as an alternative to cause–effect relations, but uses it as an analogy rather than literally. Classical and stochastic resonance can both be expressed mathematically, hence calculated. Functional resonance cannot be expressed mathematically, at least not at the present. The value of the concept is that it overcomes some important limitations of traditional safety thinking.

First, outcomes can happen without being effects of specific causes. Outcomes can be the result of coupled dynamic developments, of cycles of activity that affect each other without one necessarily being the cause and the other the effect.

Second, functional resonance can explain how non-linear effects can come about. This is especially important in cases where the signal is weak or subliminal – or in the case of stochastic resonance where the noise is weak or subliminal – since in such cases something seems to arise out of nothing, or at least out of something that usually is unnoticed.

Third, and partly as a consequence of the above, the use of resonance obviates the need to have distinct or supraliminal malfunctions. This means that adverse outcomes do not need to be explained as a result of failures, of something gone wrong, but can be seen as coming from that which happens all the time – everyday approximate performance adjustments.

# Chapter 4
# The Method: Preliminaries

The rest of this book will describe the FRAM and illustrate its use by means of three different examples. Since this description will be rather detailed, it has been distributed over several chapters. The current chapter describes the preliminaries of the method, most importantly the fact that the FRAM looks into the reasons why things go – or might go – wrong, by first describing how things go – or should go – right. Since the FRAM has been developed as a bidirectional method, some comments are also made about the practical differences between accident investigation and risk assessment.

The four chapters that follow describe each of the four steps in the method: identifying and describing the functions (Chapter 5), identifying the variability (Chapter 6), determining how variability may be combined (Chapter 7) and finally considering how the outcome of a FRAM analysis can be used to improve practice (Chapter 8). This is followed by a separate chapter that shows the use of the FRAM for three different cases.

## Looking For What Should Go Right

Accident investigation is usually about finding why something went wrong. This is based on the tradition of fault finding in technological systems, which has evolved since the Industrial Revolution began in the eighteenth century. This tradition carries with it an important assumption, namely that a (mechanical) system will work as intended unless something fails or malfunctions. In other words, failures and malfunctions are the causes of accidents, and if they can be eliminated, then accidents will not happen.

It is questionable whether this assumption is true for technical systems, but it is definitely false for socio-technical systems. The reason is that it requires that the system is fully understood, which means that work is completely specified and that there is a perfect correspondence between work-as-done and work-as-imagined. In that case we will by definition know what should have happened and can consequently limit the investigation to look for whatever might have hindered or disrupted that.

Work, however, does not take place 'mechanically', and humans and organisations are not well-behaved machines – despite the best efforts of ambitious engineers and managers to turn them into that. We can therefore not begin an investigation by assuming that we know what should have happened. On the contrary, an investigation – be it of past or future events – must begin by *establishing* what should have happened, that is, by establishing what should have gone right. This must be done for the very reason that it is generally impossible to specify work in every detail and to provide instructions that can be followed exactly as written. The basis for everyday performance is multiple approximate adjustments, and understanding how something goes right is therefore a prerequisite for understanding how it could go wrong. So instead of ascribing differences between work-as-done and work-as-imagined to errors and non-compliance, they should be seen as representing the adjustments that are necessary for everyday performance to succeed; see Figure 4.1.

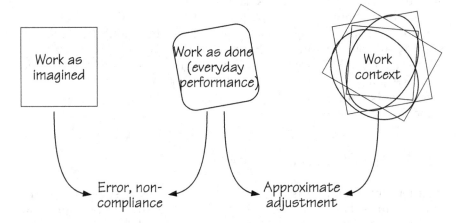

**Figure 4.1      Contrasting interpretations of work-as-done**

In the case of accident investigation, the FRAM looks for what should have gone right but did not, rather than for what went wrong – or rather, the reasons or causes why something went wrong. An investigation using the FRAM thus does not begin by looking for a cause, but by trying to understand what should have happened in the everyday case, and then used that understanding to explain why it did not happen.

Similarly, for risk assessment or looking into the future, the FRAM does not primarily look for what could go wrong, and definitely does not begin by looking for failure probabilities of individual functions or components. Instead the FRAM tries to develop a description of what should happen in the 'normal' or everyday case, and then tries to understand how the ever present performance variability may affect that either positively or negatively. ('Normal' is here used in the meaning of typical rather than in the meaning of conforming to a standard.)

The search for failure probabilities is warranted as long as it can be assumed that each component will perform in a uniform manner, and that the relations or dependencies among components remain stable or fixed. That is usually the case for mechanical or technological systems. Here components work uniformly and their relations are fixed by the design, except for the possible effects of slow degradation and maintenance. It is therefore only cases of unexpected connections that surprise us, such as sneak paths in electronic circuits or software.

The situation is completely different for socio-technical systems where neither of the two above assumptions is fulfilled. For these systems it is necessary to find an alternative approach to risk assessment. Instead of looking for whether and how a component may fail, such as in Failure Mode and Effects Analysis (FMEA), we should describe the functions that are required for the system to achieve its purpose. Following that, we should describe the potential performance variability of these functions, both individually and in combination. The combinations will often be non-linear in the sense that there is no proportionality between input and output. Specifically, a very small variation of the input may through couplings and resonance produce an exceedingly large output. This is entirely different from the essentially linear combinations of latent conditions and active failures that are described by accident theories such as Tripod (see Chapter 10). The non-linearity exists because we do not know how many steps or couplings there are between the initial event and the outcome and because the relations or dependencies among functions are not stable or fixed. The problem for socio-technical systems is actually not so much whether a specific function will work or fail as it is how several functions may become coupled. Since adverse outcomes are seen as a consequence of combinations of variability rather of than the failure of components, the issue is outside the reach of actuarial studies, at least if done in the traditional manner. What we need is a way to describe how functions can become coupled or how they can combine in unintended ways.

The advantage for risk assessment of looking for what should happen, rather than looking at a particular failure scenario – or set of failure scenarios – is that it makes the $n+1$ problem less serious. The $n+1$ problem means that if we look at $n$ failure scenarios or failure cases, then it is always possible that something may happen that has not been considered, hence that there is an *n+1* case. But if we look at a class of cases such as everyday performance rather than a (enumerable) set of specific cases, this objection carries less weight. Instead of looking at the most serious, the most recent, or the most obvious scenario – such as safety cases usually do – the FRAM looks at the most typical (and frequent) variability of everyday performance. Even here the scope of the analysis will of course be limited by time, resources and imagination, but the limitations will be less severe.

## Step 0: Recognise the Purpose of the FRAM Analysis

The first step of the FRAM is to make clear whether the analysis is tied to an event investigation that looks at something that has happened, or to a risk assessment that looks at something that may happen in the future. This step is the most obvious of the method and can in some sense be seen as superfluous. The FRAM is always used for a specific purpose and there should hopefully not be any doubt about whether it is an event investigation or a risk assessment. The reason for having a Step 0 is to set the scene for the four steps that follow, as there are slight differences in how the method is used for event investigation and for risk assessment. The four steps can be described relative to their purposes:

- The purpose of the first step is to identify the functions that are required for everyday work to succeed, by describing how something is done in detail rather than as an overall task or activity. These functions constitute the FRAM model.
- The purpose of the second step is to characterise the variability of these functions. This should include both the *potential* variability of the functions in the model, and the expected *actual* variability of the functions in an instantiation of the model.
- The purpose of the third step is to look at specific instantiations of the model to understand how the variability of the functions may become coupled and determine whether this can lead to unexpected outcomes.
- The purpose of the fourth and final step is to propose ways to manage the possible occurrences of uncontrolled performance variability that have been found by the preceding steps.

When the FRAM is used for event investigation, the first step is both helped and hindered by knowledge of the event being investigated. The knowledge helps the analysis by being a reference for the functions that must be identified. But the knowledge can also be a hindrance because it can be difficult to disregard what did happen and focus on what should have happened. The second step is affected in the same way. Knowledge of the event clearly represents one way in which the functions possibly could vary, but can also be a cognitive 'anchor' that constrains the imagination. For the third step, the instantiation to consider is clearly the event that happened, and the FRAM provides a way of looking at this instantiation as a variation of everyday performance rather than as a unique event. In the fourth step, the event can again be of help, but it must also be kept in mind that the purpose is to make sure that performance variability can be managed rather than just preventing this specific event from happening again.

When the FRAM is used for risk assessment, the first step defines the scope of analysis as well as the degree of resolution of the descriptions. It also provides a first delineation of the boundary of the system being analysed, something that becomes very important in both the second and the third steps. The second step requires a narrowing of the possible scenarios, in order to be sufficiently concrete about the instantiations and the expected actual variability. But where a traditional risk assessment starts bottom-up – by extending the initial case or set of cases, a FRAM based risk assessment is top-down, since the model is the starting point for choosing concrete scenarios. The third step continues this narrowing or specification by looking at the ways in which the consequences of performance variability may interact in the form of functional resonance. The fourth step encourages thinking about how a potentially disruptive performance variability can be detected and monitored, as well as how it can be controlled – specifically dampened.

Risk assessment using the FRAM changes from being a relatively simple calculation of aggregated failure probabilities, to become a discrete simulation of instantiations of the model. Since it is not a question of the distribution of probabilities and uncertainties for various output categories, it cannot be done automatically as in, for example, the Dynamic Logical Analytical Methodology (DYLAM), or dynamic risk modelling. A FRAM analysis acknowledges that the system constantly re-configures itself, so to speak, to match the conditions under which it functions and proposes a way in which this can be subjected to a systematic analysis.

## Comments on Chapter 4

A sneak path is an undesired function, or the inhibition of a desired function, caused not by component failures but by unanticipated cross-couplings in a system. Sneak paths were first realised in electrical circuits, where they were called Sneak Circuits. The inaugural moment was the launch of a Redstone rocket with a Mercury capsule on 21 November 1960. After a 'flight' of a few inches, lasting a mere 2 seconds, the engine cut out and the vehicle settled on the launch pad. The analysis revealed that the tail plug that connected the rocket to the launch assembly was prematurely pulled out before the control cables rather than the other way around, thereby creating a 'sneak circuit' or 'sneak path' that caused the engine to shut off. The existence of such sneak circuits made people realise that unwanted outcomes could occur even when nothing went wrong.

The distinction between work-as-imagined and work-as-done is often used in the ergonomics literature to point out that there may be a considerable difference between what people are assumed – or expected – to do and what they actually do. Work-as-imagined represents what designers, managers, regulators and authorities believe happens or should happen, whereas work-as-done represents what actually happens. Differences can either classified as non-compliances, violations, errors or as performance adjustments and improvisations, depending on how one looks at it. An early discussion of this in the context of safety is found in Turner, B. (1978), *Man-Made Disasters*, London: Wykeham.

The DYLAM was a software tool that used dynamic event trees for human error risk assessments. In the DYLAM, the analysis was based on a dynamic simulation of events, including probabilistic failures of components. This made it possible to go through many more combinations of conditions than a paper-and-pencil-based analysis. An early description of DYLAM is by Cacciabue, P.C., Amendola, A. and Cojazzi, G. (1986), 'Dynamic logical analytical methodology versus fault tree: The case of the auxiliary feedwater system of a nuclear power plant', *Nuclear Technology*, 74(2), 195.

# Chapter 5
# The Method: Identify and Describe the Functions (Step 1)

The first step of the FRAM is to identify the functions that are needed for everyday work to succeed. The purpose is to describe in detail how something is done as an everyday activity, rather than to describe it as an overall task.

It is common in human factors, in ergonomics and in psychology, to refer to tasks and/or activities. In the general literature on ergonomics the two terms are used interchangeably, and are in practice treated as synonyms. But they can also be used to describe two quite different things. In this case, a *task* describes work as it has been specified, work as it is expected or assumed to be carried out, or work- as-imagined. In contrast to that, an *activity* describes work as it is actually carried out or work-as-done. Since the FRAM deals with what actually happens – or what has happened or is likely to happen – rather than with what is or was assumed to happen, the functions refer to activities rather than to tasks. In the case of accident investigation this is hardly surprising, since the functions must represent what actually happened. But in the case of risk assessment it makes a difference whether the assessment is of the tasks, hence the idealised functions, or of the likely activities, hence the actual functions. If the idealised functions or work-as-imagined are used, then it is practically inevitable that what actually happens is considered as aberrations or deviations, hence described as 'errors', 'violations', and 'non-compliance'. If the actual functions or work-as-done are used, the need to rely on such judgemental terms disappears.

The term 'function' requires a brief explanation. The FRAM does not use the term 'function' to refer to a mathematical relation, where one quantity (the argument) completely determines another quantity (the value). The FRAM uses the term 'function' as in the goals-means relation, where a function represents the means that are necessary to achieve a goal. More generally, a function refers to the activities – or set of activities – that are required to produce a certain outcome. A function describes what people – individually or collectively – have to do in order to achieve a specific aim. A function can also refer to what an organisation does: for example, the function of an emergency room is to treat incoming patients. A function can finally refer to what a technological system does either by itself (an automated function) or in collaboration with one or more humans (an interactive function or co-agency).

## How Can the Functions Be Identified?

If the situation being investigated is something that has happened, an existing description of the event will in many cases provide much of the information needed to identify the functions. Events are often represented as a sequence of actions that leads to the outcome, for instance as an event timeline. In this case each step or action can be assumed to correspond to a function, at least as a starting point. For a prospective analysis, as in a risk assessment, a timeline is often available in the form of a specific scenario, such as a safety case. If that is not so, a different basis for identifying the functions must be found, as described below.

### Task Analysis as a Basis for Identifying Functions

The purpose of a task analysis is to describe tasks and to identify and characterise the fundamental characteristics of a specific task or set of tasks. According to the *Shorter Oxford Dictionary*, a task is 'any piece of work that has to be done', which generally is taken to mean the work or the operations that must be carried out to achieve a specific goal. Task analysis can therefore be defined as the study of what people – individually or collectively – must do in order to achieve a specific objective.

Though an overall task sometimes can be described as a 'flat' sequence of sub-tasks or operations, a task analysis usually includes sub-tasks that are organised in ways ranging from simple combinations to complex task hierarchies. The latter has been encapsulated in a commonly used approach called Hierarchical Task Analysis (HTA). An HTA decomposes tasks into sub-tasks and repeats this process until a level of elementary tasks has been reached. Each sub-task can be specified in terms of its goal, the operations required to attain the goal, and the criteria that mark the attainment of the goal. The relationship between a set of sub-tasks and the superordinate task is governed by plans, procedures, selection rules or scheduling principles.

A simple example of a HTA can be seen in Figure 5.1. It represents the steps needed to get money from a bank account using an Automated Teller Machine (ATM). The HTA includes an upper level that defines the order of the main tasks, and a lower level of sub-task where each corresponds to a specific operation or set of operations. It is clearly possible to describe each of the sub-tasks in further detail, and to do that repeatedly. This both means that there is no absolute 'lowest' level of description and also raises the question of what the elementary tasks are, that is, what the stop rule of the HTA is.

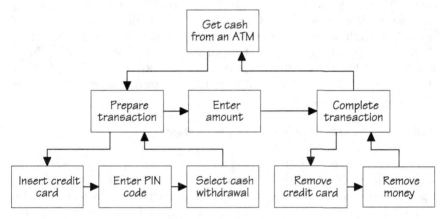

**Figure 5.1    A Hierarchical Task Analysis of cash withdrawal from an ATM**

As an example of how a sub-task always can be expanded further, consider the 'Enter PIN code' in Figure 5.1. Most people would readily accept that this is a meaningful elementary task because everyone 'knows' how to enter a PIN code. It is nevertheless possible to describe this in further detail, for instance as follows: [<press on the key for the first digit>, <press on the key for the second digit>, <press on the key for the third digit>, <press on the key for the fourth digit>, <press on the ACCEPT key>]. This illustrates that 'Enter PIN code' is not necessarily an elementary task, and also suggests that each of the four other sub-tasks may be subjected to further decomposition.

A FRAM analysis of the cash withdrawal situation could use the HTA as a basis for identifying the following functions: [<Enter amount>, <Enter PIN code>, <Get cash from an ATM>, <Insert credit card>, <Remove cash>, <Remove credit card>, <Select cash withdrawal>]. While the HTA describes the temporal relationships between functions (Figure 5.1), the order in which the functions take place is not represented explicitly in the FRAM model. The seven functions have therefore been listed alphabetically rather than in the order in which they occurred in the HTA diagram. After the functions have been identified, the next step is to characterise each by up to six aspects. The descriptions of the aspects define the potential relations among functions, which need not be the same as those found by the HTA.

An alternative to task analyses such as HTA is to use a goals-means analysis. Whereas task analysis was not used systematically before the beginning of the twentieth century, the principle of goals-means analysis can be traced back at least to Aristotle. This is, of course, not really surprising since decomposing wholes into parts has been a trusted method in human thinking as far back as we can imagine. And although the physical nature of work has changed throughout history, not least after the Second Industrial Revolution, the logical descriptions of how to do things are largely independent of the level of technology available.

## Task Descriptions versus Function Descriptions

Task descriptions usually have a clear structure. A task description not only identifies and characterises the functions that together make up the tasks, but also how they relate to each other in terms of the order of execution. It is therefore obvious where the description should begin and where it should end. The basic organising principle is that each step of the task brings the acting agent closer to the stated goal. This is therefore also the principle of decomposition.

Most existing approaches to task analysis, such as an event timeline or an HTA, describes the functions and how they are related. The relation can be temporal, as in a timeline, or be in terms of control or a superordinate-subordinate relation, as in an HTA. In either case it is important to distinguish clearly between a description of the functions and a description of the task of which the functions are parts.

A FRAM analysis can be carried out in the same manner as a task analysis, starting at the first step or function and progressing from that. But a FRAM analysis can also begin by any function, for instance one that appears to be essential for the task or scenario considered. By describing each function further (see below), the FRAM will iteratively identify any other functions that may be needed to provide a complete and consistent description of the task or activity.

## Boxes and Arrows

It is common to represent actions, events, processes or functions as flowcharts, where boxes or other shapes represent the process elements and connectors, lines or arrows, represent the relations between process elements. The flowchart has been used to represent work flow since the early 1920s, and was in the 1940s adopted to illustrate computer algorithms. Flowcharts are today the standard means to explain the details of how a system functions.

One weakness of flowcharts is that the arrows or lines connecting the process elements usually lack an explicit meaning or interpretation. They represent some kind of influence, but exactly which is 'intuitively' understood or taken for granted, and therefore not explained explicitly. In a task analysis diagram, such as in Figure 5.1, the lines represent a transfer of activity or control in the sense that if task A has been carried out, then task B follows. In other cases, such as process diagrams, the lines can represent the transportation of something (information, energy or matter) between two functions or system components.

A simple example of a flowchart is shown in Figure 5.2. Here, the link ① represents a temporal relation or perhaps the transmission of information. 'Process' has three inputs and one output. The link ④ is the transmission of something that presumably is used by the process, whereas link ③ is a logical link – or possibly information. The link ② is a transfer of control, in the sense that whenever a process cycle has been completed, a test is carried out to determine whether the job is finished. If the answer is no, then the process is repeated; if the answer is yes, then the process ends.

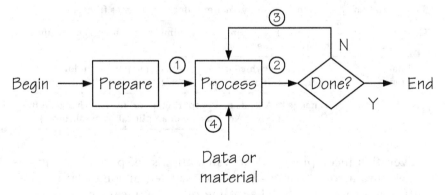

**Figure 5.2    A simple flowchart**

Even this simple flow diagram contains at least three different types of links, although none of them is defined explicitly. Instead it is simply assumed, undoubtedly with some justification, that users understand what the arrows mean and therefore are able to supply the missing information. Another characteristic feature of Figure 5.2 is that the order of the three main functions is clearly marked. The three functions will therefore always take place in the order <Prepare> – <Process> – <Done?>, with a possible repetition of the last two steps.

In contrast to that, the FRAM simply recognises that there are three functions in Figure 5.2, and represents them as the unordered set [<Done?>, <Prepare>, <Process>]. The FRAM does not explicitly link or relate the functions to each other, for example, as boxes connected by lines. Instead each function is characterised in terms of some aspects, which in turn make it possible to derive the potential relations (couplings) between functions.

Considering the multitude of diagrammatic representations of relations – from family trees to flow charts – the commonly used meanings of lines and arrows seem to be the following.

**Table 5.1      Commonly used meanings of arrows and lines**

| Relation | Explanation – example |
|----------|------------------------|
| Causal | A is the cause of B – and B is the effect of A. |
| Temporal | A happens before B, or A is a precursor to B. This can also be seen as representing 'flow,' hence is the basis of flowcharts. |
| Abstraction-concretisation | A is a parent or superset of B, or B is a child or subset of A. This is the basis of most taxonomies. |
| Transportation | Something (energy, matter, information) goes from A to B. |
| Control | A controls B. If B comprises a number of actions, A determines the order of execution. |
| Ordinal relations | A is before B (which usually means a temporal or a flow relation, but it may not have been specified). |
| Change | A change in A leads to a proportional (or inverse) change in B. (This can also be seen as a case of amplification-weakening.) |

Even this incomplete and unsystematic list of possible meanings of lines and arrows demonstrates a clear need of some clarification, lest the interpretation of what a line or arrow means can be left to the users. The FRAM recognises this issue and provides a way of describing five specific relations among functions.

## Making Arrows and Lines Meaningful

Since the arrows or lines used in flowcharts always do have a meaning – or a semantics – it is a good idea to make it explicit. This will not only make the representation more powerful, but also reduce the possibility of misunderstandings.

The need to define the meaning of the links explicitly has, of course, been recognised a number of times and addressed by several methods. A good example of that is the Structured Analysis and Design Technique (SADT). This technique was developed in software engineering to describe systems as a hierarchy of functions. The basic component of the SADT is an activity. But unlike the general flowcharts, each activity is described by four clearly defined links:

- *Inputs* to the activity represent data or consumables that are needed by the activity. Inputs enter from the left.
- *Outputs* from the activity represent the data or products that are produced by the activity. Outputs leave from the right.
- *Controls* represent commands that influence the execution of an activity but are not consumed. Controls enter from the top.
- Finally, *mechanisms* represent the means that are necessary to accomplish the activity, such as components or tools. Mechanisms enter from the bottom.

In an SADT diagram, the meaning of a link is shown by how it enters or leaves an action (box); see Figure 5.3.

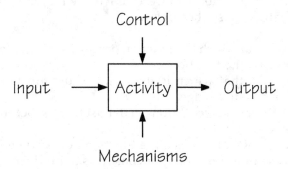

**Figure 5.3     An SADT activity**

In the FRAM, links or couplings among functions are equally well defined. There are, however, no arrows since a FRAM model is the description of the functions rather than the diagram. Neither is the meaning of the links associated with a specific position or direction.

Even though there is a practical tradition in drawing FRAM diagrams (see Figure 5.4), links can in principle be placed anywhere. This is because functions need not be ordered or positioned in any special way in the diagram, for example, left-to-right or top-to-bottom.

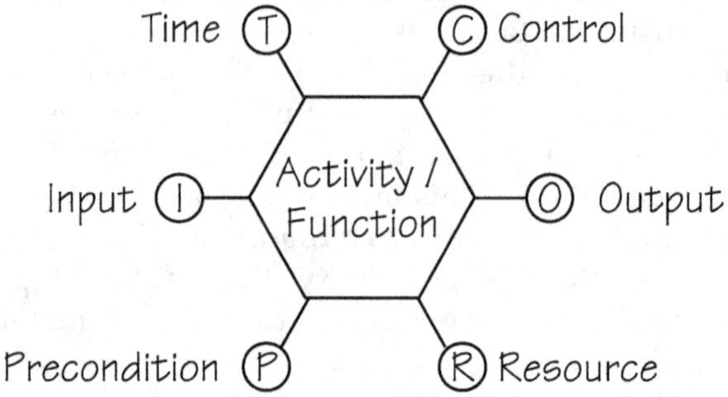

Time (T)                    (C) Control

                Activity /
Input (I)       Function        (O) Output

Precondition (P)        (R) Resource

**Figure 5.4      A hexagon representing a function**

The FRAM also makes a clear distinction between the functions themselves and how they can be related or coupled. The first part of Step 1 has addressed how the functions could be identified. The second part of Step 1 will explain how a characterisation of the functions can be used to determine how they can be coupled, potentially and actually.

## The Six Aspects

In the FRAM, a function can be characterised by the six different aspects or features described below:

- Input (I): that which the function processes or transforms or that which starts the function.
- Output (O): that which is the result of the function, either an entity or a state change.
- Preconditions (P): conditions that must be exist before a function can be carried out.
- Resources (R): that which the function needs when it is carried out (Execution Condition) or consumes to produce the Output.
- Time (T): temporal constraints affecting the function (with regard to starting time, finishing time or duration).
- Control (C): how the function is monitored or controlled.

A FRAM function is represented graphically by a hexagon, where each corner or vertex corresponds to an aspect, as shown in Figure 5.4. Since each corner is labelled, a graphical rendering of a FRAM model does not require the hexagons always to be oriented as shown in the figure, although it may be convenient to do so. The orientation of the hexagons is anyway of minor interest since a FRAM analysis should be based on the verbal description rather than the graphical representation. (This point will be elaborated below.)

Each of the six aspects is explained in further detail in the following sections. In order to reduce possible ambiguities, each aspect will be written with a capital first letter. In this way Input refers to the aspect of a function that represents the Input as defined below, whereas input refers to the four other aspects of a function that may receive the Output from upstream functions.

## Input

The Input to a function is traditionally defined as that which is used or transformed by the function to produce the Output. The Input can represent matter, energy or information. This definition corresponds to the normal use of the term in flowcharts, Process and Instrumentation Diagrams, process maps, logical circuits and so on. There is, however, another meaning that is just as important for the FRAM, namely the Input as that which activates or starts a function.

The Input in this sense may be a clearance or an instruction to begin doing something, which of course must be detected and recognised by the function. While this nominally can be seen as being data, the Input primarily serves as a signal that a function can begin. Technically speaking, the Input represents a change in the state of the environment, just as if the Input was matter or energy. If, for instance, the Input is a clearance or work-order, then the state before the Input arrived was [Presence of clearance: False] while the state after the Input has arrived is [Presence of clearance: True]. It is when the state changes from [False] to [True], that the function can begin. This change of state may be concrete such as a piece of paper in an in-tray, an order to a chef, a patient arriving to an emergency room, or the fact that there are no more cars waiting to be loaded onto a ferry. The change of state may also be symbolic or abstract, such as a threat advisory level or the WHO alert level for avian influenza.

The role of the Input as a signal that marks the beginning of a function suggests how the variability of functions can arise. The detection threshold can either be too high or too low, the Input can be misinterpreted or mistaken for something else, and so on. More will be said about this in Chapter 6.

It is common in flowcharts and similar diagrams to represent an input-output relation by a line or an arrow. But that is not the case in the FRAM. Instead, the Input – or Inputs, since there may be more than one – is simply an aspect of a function. The Input to a function must clearly correspond to an Output from at least one other function. Indeed, the FRAM makes sure that such a correspondence exists. This correspondence represents the potential but not the actual couplings among functions. The actual couplings are those that can realistically be assumed to exist for a given set of conditions (called an instantiation). While the actual couplings always will be a subset of the potential couplings, they may be different from the couplings that were intended by the system design.

While nearly all functions will have Inputs, it is not always required. In the FRAM, designated foreground functions must have defined Inputs, while designated background functions need not have. The difference between foreground and background functions will be explained later in this chapter.

*Output*

The Output from a function is the result of what the function does, for instance by processing the Input. The Output can therefore represent matter, energy or information – the latter as a command issued or the outcome of a decision. The Output can be seen as representing a change of state – of the system or of one or more output parameters. For the simple example of getting money from an ATM, the Output can either be described as the money that the person now has (in which case the Output can be said to be material or matter), or a change of state, for example, [Person has money: True]. The Output can, of course, also represent the signal that starts a downstream function.

The Output is important to explain how variability can propagate in a system. If a function varies, for one reason or another, then it must be assumed that the Output will also vary in some way. (Indeed, if the Output does not vary, then there is no reason to consider the variability of the function.) We have already seen how the variability of the Input to a function may lead to variability in the way the function is carried out. Since the Output from a function will be inputs to other functions, the variability of these inputs may lead to variability in the other functions, hence to increased variability of their Output, and so on. Note, however, that the situation need not always develop for the worse. It is quite possible that a function is able to compensate for or dampen the variability of the Input it receives, so that its Output remains unaffected. Indeed, this is one way to understand the resilience of a system.

## Precondition

A function can in many cases not begin before one or more Preconditions have been established. These Preconditions can be understood as system states that must be [True], or as conditions that ought to be verified before a function is carried out. A good example is the case of the *Herald of Free Enterprise* (Chapter 9) where there were three Preconditions before the ferry could leave port: that the moorings had been dropped, that the ferry was in trim and that the bow ports were closed. In the case of the *Herald*, the first Precondition was fulfilled but neither of the others were, nor were they even checked.

If the performance of a function does not vary, it means that existing Preconditions will be checked as expected, if so required. But it is a general experience that this not always is the case – to the extent that a disregard of Preconditions is the rule rather than the exception. In the *Herald of Free Enterprise* case, the captain applied a 'negative reporting' principle, meaning that if no one reported that a condition was false (for example, that the bow ports were open), then it could (safely) be assumed to be true, in this case meaning that the bow ports were closed.

In addition to providing a way by which functions can be coupled, Preconditions also help to find the functions that are necessary for a FRAM model to be complete. The simple principle is that any Precondition defined for a function should correspond to an Output from another function. This is exactly the same as for the Input, and also applies to the remaining three aspects (Resources, Control, Time).

One issue that occurs in the use of the FRAM is the difference between Input and Precondition. To explain this difference, consider the example of an aircraft which taxies from the gate to the runway. When it has reached the runway it must wait at the hold-short line until it receives a clearance from ATM. The aircraft cannot take-off without that (at least if the pilots follow the procedures). This clearance serves as an Input because the take-off starts when (or shortly after) the clearance has been received. If the clearance was characterised as a Precondition, meaning that the aircraft had to wait for it before it could take off, then there would have to be some other (Input) signal that started the function. It therefore makes sense to treat the clearance as an Input. In the same example, pilots must also complete the take-off checklist before they can take off. However, having completed the checklist does not in itself permit them to take off; they still have to wait for the clearance from ATM. The completion of the take-off checklist is therefore a Precondition for taking off, but not an Input.

In summary, a Precondition is a state that must be true before a function is carried out, but the Precondition does not itself constitute a signal that starts the function. An Input, on the other hand, can activate a function. This simple rule can be used to determine whether something should be described as an Input or as a Precondition. It is, however, not critical for a FRAM analysis whether something is labelled an Input or a Precondition, as long as it is included in the model in one way or another.

## Resources (Execution Conditions)

A Resource is something that is needed or consumed while a function is carried out. A Resource can represent matter, energy, information, competence, software, tools, manpower and so on. Time can, in principle, also be considered as a Resource, but since Time has a special status it is treated as a separate aspect. Since some Resources are consumed while the function is carried out and others are not, it is useful to distinguish between *Resources* on the one hand and *Execution Conditions* on the other. The difference is that while a Resource is consumed by a function, so that there will be less of it as time goes by, an Execution Condition is not consumed but only needs to be available or exist while a function is active. The difference between a Precondition and an Execution Condition is that the former is only required before the function starts, but not while it is carried out.

Resources typically represent material and/or energy. More generally they can be seen as a property of the system that changes in some specific manner while the function is carried out. Since the change normally means that the Resource is spent, it is necessary that it is either maintained or replenished. A Resource for a physical process could be, for example, the pressure or temperature of a fluid, a supply of fuel or necessary raw materials and components. For an organisational process it could be capital or a revenue stream (a stable or growing number of customers) and so on. There must either be sufficient Resources to allow the function to continue for the required time (for example, the fuel carried by an airplane), or there must be some way of replenishing them (for example, getting a loan to keep a business running). Providing the Resources in the first place, or replenishing them as they are being used, are of course both functions that must be described in their own right.

Execution Conditions typically represent tools and technology, the ambient working conditions, data or information, but can also represent something less tangible such as competence or skills. Execution Conditions must be present for the function to be carried out but are not consumed or diminished as a consequence of that. A hammer, or a computer, may be necessary for an activity, but neither the hammer nor the computer changes during the activity. (If they did, for example, through 'wear and tear' for a very extensive activity, then they should instead be considered as Resources.) Information (knowledge) and competence may also be necessary to do something, but neither are 'used or diminished' because of that – in some cases they may actually increase!

The practical criterion to determine whether something should be treated as a Resource or an Execution Condition is whether there will be less of it as the function is carried out or whether it can be assumed to be stable or constant – that is, whether changes to its quantity or quality are negligible. In the former case it should be considered a Resource, and in the latter an Execution Condition.

Resources and Execution Conditions are very often taken for granted in the description of functions, hence not explicitly defined. Examples of such obvious Execution Conditions are oxygen (so people can breathe), light (so people can see) and so on. If we can assume that a Resource or an Execution Condition generally will be available and not vary significantly, then it need not be included in the model. This is actually a practical necessity, since the model otherwise would have to encompass the whole world. There is, of course, always the constant sense of unease that the assumptions about the availability of a Resource may be wrong. But this is the unavoidable ETTO condition of any analysis.

## Control

Control, or control input, is that which supervises or regulates a function so that it results in the desired Output. Control can be a plan, a schedule, a procedure, a set of guidelines or instructions, a program (an algorithm), a 'measure and correct' functionality and so on.

All functions must have some kind of Control, either built into them or provided externally. Consider, for instance, the 'Enter amount' function in Figure 5.1. In the description, it represents a single function, but it can be analysed in further detail as a sequence of entering individual digits. Yet even entering a digit is not a simple perceptual-motor activity but requires some kind of Control. In a FRAM analysis it is, however, sufficient to describe the Control that has an external origin. This could be, for instance, a procedure developed by the design department, a guideline developed by a regulatory authority or by company management, the detailed instruction for how to assemble (or disassemble) something, and so on. In other words, the Control represents the explicit instructions (in a wide sense of the word) that relate to the function, and that (presumably) are required to bring about the goal. These can either be the standard or regular instructions, or special instructions for the specific situation, that is, an ad hoc guideline or a set of exceptions. For humans, Control can also be their expectations to the work, what they should look out for, and so on. In other words, Control is that which people refer to while performing a function, either as something they know (remember or have learned), or as something that has an external (symbolic) representation.

A different type of Control is social control and/or expectations. Social control can either be *external*, for example, the expectations of others, for instance the company or management, or *internal*, such as when we plan do to some work and make clear to ourselves when and how to do it. Internal social control can be seen as a kind of self-regulation. External social control will typically be assigned to a background function, *q.v.*

## Time

The *Time* aspect of a function represents the various ways in which Time can affect how a function is carried out.

Time, or rather temporal relations, can be seen as a form of control. One example of that is when Time represents the sequencing conditions. (This can also be seen as a kind of Control.) A function may, for instance, have to be carried out (or be completed) before another function, after another function, or overlapping with – parallel to – another function. Time may also relate to a function alone, seen in relation to either clock time or elapsed time. The sequence conditions can be expressed as relative timing conditions using four self-explaining temporal relations:

1. Earliest Starting Time (EST)
2. Latest Starting Time (LST)
3. Earliest Finishing Time (EFT)
4. Latest Finishing Time (LFT)

It is clear that these timing conditions need only refer to an ordinal rather than an interval scale. As such they neither require estimates of duration nor references to measured time (absolute or elapsed). Even ordinal timing conditions are, however, extremely useful to provide further details for the description of Preconditions or Execution Conditions. A timing condition could, for instance, be that a function must begin after another has ended, or that two functions must be carried out simultaneously.

Time can also be seen as representing a Resource, such as when something must be completed before a certain point in time (LFT), or within a certain duration (as when a report must be produced in less than a week). Time can, of course, also be seen as a Precondition, for example, that a function must not begin before midnight, or that it must not begin before another function has been completed. Yet rather than having Time as a part of any of the three aspects of a function – or indeed, the four since it conceivably could also be considered as an Input – it seems reasonable to recognise its special status by having it as an aspect in its own right.

## Issues in the Description of Functions

The preceding sections have explained why the FRAM characterises functions by six different aspects, how the aspects should be understood, and how they should be described. This way of characterising a function makes it possible to understand how performance can vary (Chapter 6), and how performance variability can combine to produce unanticipated and out-of-scale consequences (Chapter 7). With the risk of making the reader feel nauseous, functions do not simply work or fail, but rather vary in how they are carried out. The variability of a function may give rise to variability in the Output from a function. Outcomes of events – normal or excessive, desired or undesired – can consequently be understood in terms of the variability in the Output from functions, hence as a propagation of the variability of functions, instead of as the effects of identifiable failures.

The description of functions as well as of the aspects of functions, is, however, not always a simple matter. The concepts and the terminology provided by the FRAM are meant to be of use in developing an understanding of how events may develop – or of how events have developed – but are not by themselves sufficient to provide the understanding. It is impossible to make a safety assessment or a safety analysis in the absence of subject matter expertise, with the possible exception of systems that are so simple that a safety analysis is not needed anyway. An analysis is not an algorithmic process but rather the gradual development of a mutual understanding among a team of experts working together. The practical value of a method is to provide a common frame of reference to define which data are needed, to help relate findings and to express conclusions. A method will inevitably constrain the analysis in some way, and different methods do it in different ways. A method should therefore always be chosen to match the needs of an investigation, rather than be applied in a wholesale fashion.

## When Does A Function Start and Stop?

Since the FRAM describes functions, it is important to be able to define when a function starts, and equally important, when it stops. Referring to the descriptions of the six aspects, a function can start or begin for the following reasons:

- A change in the Input, corresponding to a change in system state, typically described as a condition that becomes true. This can either be a direct signal or a recognition that something has changed.
- A Time relation, for example, as when a function is supposed to begin at a specific time (relative or absolute).
- A Control relation, for instance a step in a procedure such as 'Start the pump when the temperature has reached 220°C'.
- A change in a Precondition, or rather that the system state specified by the Precondition changes, for instance that a checklist has been completed.

A function can cease or stop for the first three reasons, but also for a fourth, namely if a Resource or an Execution Condition no longer is available or present.

The starting, and stopping, of a function may be under-specified or ambiguous, which means that there may be 'false' starts as well as 'false' stops. Time pressure may, for instance, override a Precondition, so that a function begins even though the situation was not quite ready, that is, even if the Preconditions were not fulfilled. Or if too much attention is paid to the text of a procedure, changes to the Input may be missed or neglected. Finally, a Precondition may lead to a false start, for example, if the person (or team) pays too much attention to the Precondition and too little attention to the Input (unbalanced priorities). In the other direction, a function may be carried out for too long because a stop condition is overlooked or because too many interruptions degrade the level of control.

## Functions and States

Functions always describe something that can be done or is being done. The description is thus always of an activity, which in linguistic terms means that it must contain a verb phrase. The purpose of a function is to produce something or to bring about a state change. The result of a function, the Output, is always a description of a (system) state, or of a condition.

In the *Herald* example, <close the bow doors> is a function, while [bow doors are closed] is a state, that is, the Output from the function. In the financial example (Chapter 9), <get a credit risk assessment> or <assess the risk of company X> is a function, while [credit risk is acceptable] is a state. The credit risk is the result, while the checking or interrogation itself is a function.

While this rule may seem rather trivial, it is important to distinguish between a noun phrase and a verb phrase when the functions are described. The general rule is simple enough: the name of a function must be a verb phrase, that is, describing something that is being done, while the name of any aspect must be a noun phrase, that is, describing a state.

## Boxes and Arrows Revisited

As discussed above, the arrows or lines that connect the elements of a flowchart do not normally have a well-defined semantic. In the FRAM, on the other hand, the possible couplings among functions are clearly defined. The 'elements' are also well-defined, since they are the functions – currently shown as hexagons.

Having said that, it is important to emphasise that a FRAM model is the textual description of the functions and their aspects. The graphical representation is useful for communication, but a FRAM analysis should be based on the textual descriptions rather than the graphics. The obvious reason for this is that the graphics very quickly becomes unwieldy as each function has (potentially) six types of connection rather than two. Another reason is that a FRAM model describes the potential couplings, rather than the actual couplings, which are relevant (or limited by) a specific scenario or instantiation. In consequence of this, the representation of a function should not be the hexagon shown in Figure 5.4 but the simple form shown in Table 5.2.

In a FRAM model there is no a priori order or sequence of the functions. The model consists of a set of function descriptions, using the format shown above. The functions are potentially coupled if they have common descriptions of the aspects. For instance, if something is described as the Output for function A, and also as a Precondition for function B, then A and B are potentially coupled. The logic of a Precondition also means that A nominally should be carried out before B. However, in a given instantiation of the model, function B may vary and therefore start too early, possibly even earlier than function A. So even if the Output from one function can be matched to the Precondition of another, that does not mean that one will always be carried out before the other. In fact, the main point of a FRAM-based

**Table 5.2     A FRAM frame**

| Name of function | |
| --- | --- |
| **Aspect** | **Description of aspect** |
| Input | |
| Output | |
| Precondition | |
| Resource | |
| Control | |
| Time | |

analysis is to understand how the potential couplings of a model can become actual couplings in an instantiation of the model.

### Upstream and Downstream Functions

In the FRAM model, the potential couplings are defined by means of the description of the aspects of the various functions. But since the FRAM model describes the functions and their characteristics, rather than how the functions are organised for specific conditions (either a situation that has happened or a situation that is considered for the future), it is not possible to say whether one function will be carried out prior to or after another function. This is what is achieved by creating one – or more – instantiations of the model.

In the instantiations, detailed information about a specific situation or scenario is used to create an instance or a specific example of the model (hence, the term *instantiation*). In the case of a risk assessment, an instantiation can be seen as the result of applying an 'if, then' exercise to the model. This will lead to a temporal organisation of the functions as they are likely to unfold (or become activated) in the scenario. Note again, that the instantiation may differ from the intended sequence or order of functions because the performance variability affects the actual couplings. In the case of an event investigation, an instantiation of the model can be used to explain how the specific set of conditions and circumstances that existed at the time led to the observed outcomes.

It is only possible to consider functions in their temporal and causal relations when an instantiation has been produced. This is why functions that – in the instantiation – happen before other functions, and which therefore may affect them, are called upstream functions.

Similarly, functions that happen after other functions, and which therefore may be (or are) affected by them, are called downstream functions.

Consider an instantiation of a FRAM model where function A is carried out before function B, and function B is carried out before function C. Relative to function A, function B is a downstream function. But relative to function C, function B is an upstream function, just as function A is an upstream function to B. Any given function may thus at one time be described as a downstream function and at another time as an upstream function. The terms, however, only have meaning for an instantiation of a model, but not for the model as such.

## Foreground Functions, Background Functions and Performance Shaping Factors

While upstream and downstream functions refer to the temporal relations among functions in instantiations of a FRAM model, foreground and background refer to the relative importance of a function in the model.

It has been common practice in human reliability assessment to invoke the presence of so-called Performance Shaping Factors (PSF) to describe the conditions that influence the events being studied, not least in the case of the 'human error' rate. The PSFs are a thus way of introducing variability – at least in the sense of having an uncertainty interval around the calculated or estimated failure probability. In this tradition, the PSFs represent the background factors (the context or the environment), that play a role for the phenomenon being studied.

The rationale for the PSFs was the common focus of Human Reliability Assessment (HRA), namely human performance and more specifically 'human error'. When the focus is fixed, then it makes sense to describe the rest as the context. The PSFs express the simple fact that human performance, and indeed the performance of most systems, depends on the conditions in which the performance takes place.

Performance (shaping) conditions have typically been referred to as factors, that is, as characteristic traits of the environment or the context. The factors can be assigned various values that can be used to determine their impact on performance. This can be done qualitatively, but has mostly been done quantitatively. Few approaches, however, try to account for the possible relations between the factors, that is, for how performance conditions may depend on each other. Indeed, in most cases, such as the classical PSF, they are assumed to be independent, which conveniently allows the combined effect to be determined by simply adding them together.

It is, however, more reasonable to assume that performance shaping factors or performance conditions are mutually dependent, and that this may affect their impact. But that means that the PSFs cease being factors in the traditional meaning of the term and instead become the outcome or result of something which in itself can vary. In the FRAM terminology, the factors are functions themselves, or rather the result of functions, and they are consequently similar to the (foreground) functions they are assumed to influence. The developments in safety thinking and risk assessment directly support this view. The general trend since the 1970s has been to widen the focus going from a preoccupation with technical failures to include first 'human error' and later organisational failures, organisational culture and the like. But when the focus changes from human performance to organisational performance, then the nature of the performance conditions or performance shaping factors must also change. When the focus is on human performance, the performance conditions can include the organisation. But when the organisation becomes the focus of study, then the performance conditions must be something else, such as the economic climate, the political climate and so on.

The FRAM makes this redefinition of performance conditions superfluous by representing them as functions. The foreground functions denote that which is being analysed or assessed, that is, the focus of the investigation (for instance how a ward in a hospital functions). The background functions denote that which affects the foreground functions being studied. In other words, the background functions constitute the context or working environment. These background functions should clearly be described in exactly the same way as the functions that are being studied, although possibly with less detail.

As we have already seen in the description of the six aspects of functions, there may be a need to define functions to provide the Inputs, Preconditions, Resources/Execution Conditions, Control, and/or timing of other (downstream) functions. In many cases it can be assumed that functions defined in this way do not vary significantly in the situation, which means that there is no need to model or describe them in detail. The differences between foreground and background functions thus typically refer to the timescale or the dynamics of the functions. Using the analogy of the visual figure-ground phenomenon, we consider the performance shaping conditions as background functions (the perceptual ground). The analogy means that just as the figure only can be seen relative to the ground, so can the foreground functions only be understood relative to the background functions. And just as figure and ground can switch so can a function sometimes be foreground and sometimes background. In the FRAM this switch happens if the assumption about the limited variability of a background function turns out to be incorrect. This means that the background function now must be included in the study – for example, added to the set of foreground functions. This may in turn lead to the specification of additional background functions and so on, until the model has been completed.

*Other Issues in the Description of Functions*

Before concluding this section, a number of other issues require consideration. This will be done briefly here as a more thorough discussion must wait for another time and/or place:

- Degree of elaboration or detail of the descriptions. In a FRAM model, the descriptions of the various functions can have different degrees of elaboration. A FRAM model will typically comprise functions described in detail, for instance the designated foreground functions. If there is reason to assume that there can be significant variability in the performance of a foreground function, then it is makes sense to develop the description in further detail, for instance by replacing one function by two or more other functions (decomposition or increasing resolution). Similarly, if there is no reason to assume that there is significant variability for a set or group of functions, then it makes sense to collapse or aggregate them into a single function – knowing that the change can always be reversed.
- System boundary (stop rule). This is the issue of how far the analysis should continue, that is, when a FRAM model is (reasonably) complete. As the description of the FRAM already has shown, the analysis may often go beyond the boundaries of the system as initially defined. This could be because the description of the aspects of a function makes it necessary to include additional functions in the model. Or it could be because some background functions may be found to vary and thereby affect designated foreground functions, in which case they should be treated as foreground functions and the boundary extended correspondingly. The FRAM thus has a semi-explicit stop-rule built in, namely that the analysis should continue until there is no unexplained (or unexplainable) variability of functions – which is the same as saying that one has reached a set of functions which can be assumed to be stable rather than variable. But the boundary is relative rather than absolute, and refers to functional characteristics rather than physical characteristics.

- Model consistency and completeness. The consistency of a FRAM model requires that aspects that refer to the same state used the same names or labels. If, for instance, one aspect is named [Harbour stations are manned] and another [Harbour stations have been manned], then it should be ascertained whether this actually refers to the same aspect. If so, the names should be made identical. If not, the names should be made more distinct. The completeness can be checked by making sure that all aspects described for one function are included in the aspects described for other functions. In other words, no aspects should occur for one function only. This, of course, requires that the consistency has been checked first.

- Another issue is whether all aspects of a function must be described. The answer is clearly that this is not necessary. For all foreground functions it is necessary to describe, at least, the Input(s) and the Output(s). (For some background functions it may be sufficient to describe the Outputs, if they serve as placeholder functions.) On a case-by-case, or function-by-function, basis the analysis team must decide whether it is necessary (or indeed possible) to describe any of the other aspects. This judgement is typically rather easy to make in a concrete situation. It is furthermore something that can quickly be amended when the need arises: one can simply add a description of an aspect. Notice that this is possible mainly because the analysis is based on the textual description of the FRAM node, rather than on the graphical. Making changes to the graphical description can be very laborious, time-consuming and confusing. Making an amendment of the textual description is much easier, but must be followed by a check for consistency and completeness.

- A final issue is whether all aspects are equally important, traditionally expressed as whether they should all be given the same weight. The answer is clearly that they are not. This is simply because the functions can be very different, hence that they may depend on aspects in different ways. The importance of an aspect can only be understood relative to a specific function. For some Time may be important, while for others Time may be irrelevant. The same goes for the other aspects.

## Comments on Chapter 5

A Hierarchical Task Analysis (HTA) basically decomposes tasks into sub-tasks and repeats this process until a level of elementary tasks has been reached. The overall aim is to describe the task in sufficient detail, where the level of resolution required depends on the specific purposes, for example, interaction design, training requirements, interface design, risk analysis and so on. Descriptions of HTA can be found in many handbooks, as well as on the web. The original reference is Annett, J. and Duncan, K. D. (1967), 'Task analysis and training design', *Occupational Psychology*, 41, 211–221.

As an aside, the now-ubiquitous flowchart can be traced back to the 'process charts' of 'flow process charts' created by Frank Gilbreth in 1921 to illustrate the flow of work processes or, in other words, an early form of task analysis. In 1947 it became a basic technique of representation for computer programming, and has since spread to all kinds of professional activities.

The Structured Analysis and Design Technique (SADT) was described by Ross, D. T. (1977), 'Structured analysis (SA): A language for communicating ideas', *IEEE Transactions on Software Engineering*, SE-3(1), 16–34. The technique was developed in the field of software engineering but has also been used to model the functions of a power plant in a safety analysis.

The characterisation of the various temporal relations first appeared in Allen, J. (1983), 'Maintaining knowledge about temporal intervals', *Communication of the ACM*, 26, 832–843. It remains a highly readable paper.

The realisation that human performance depends on the conditions can be found in the origins of scientific psychology, around 1879, but is certainly much older than that. It was an important part of the Technique for Human Error Rate Prediction (THERP), which arguably is the best known first-generation HRA method. THERP was described in Swain, A. D. and Guttmann, H. E. (1983), *Handbook of Human Reliability Analysis with Emphasis on Nuclear Power Plant Applications*, NUREG/CR-1278, USNRC.

# Chapter 6
# The Method: The Identification of Variability (Step 2)

The purpose of the second step is to characterise the variability of the functions that constitute the FRAM model. This step should address both the *potential* variability, referring to the model, and the expected *actual* variability, referring to an instantiation of the model; see Chapter 5. In either case both foreground and background functions should be considered. The identification of variability should take into account both everyday – or 'normal' – variability and possible cases of unusual – 'out of range' – variability. The latter can, of course, be either harmful or beneficial, corresponding to traditional distinction between threats and opportunities.

In the FRAM, the characterisation of performance variability is needed to understand how functions can become coupled and how this can lead to unexpected outcomes. This means that the analysis really is looking more at the variability of the Output from functions than at the variability of functions as such. To put it differently, if the Output from a function does not vary even though the function itself varies, then the variability of the function is in principle of no interest. But if the Output from a function varies, then the variability of the function becomes important because it is that which determines the quality, hence the variability, of the Output.

Looked at from above, any system can be represented by a single function. The function of a hospital, for instance, is to treat patients, the function of an Air Navigation Service Provider is to prevent that aircraft come too close, the function of a ferry is to transport people and goods from one point to another, and so on. Seen from this perspective most systems show little variability – as indeed they should. This does not mean that a system is not variable if looked at in greater detail, that is, if it is decomposed. It rather means that the system is able to absorb the variability of the more detailed functions so that it does not show in the more abstract description.

Consider, for example, a car assembly plant that functions well according to the principles of lean manufacturing or Kaizen. Such an assembly plant will show little variability as a whole in the sense that the output, the vehicles it produces, all will be of consistently high quality. But looked at in greater detail there will be variability. Indeed, one of the basic ideas of lean manufacturing is that attention should be paid to the variability of the detailed processes so that it can be identified and corrected locally before the consequences manifest themselves in the aggregated product. In this way variability is absorbed or dampened internally, and a FRAM analysis can show how this happens. That is, in fact, the purpose of Step Four of the method. Having said that, there may be functions that produce a near constant (reliable) Output even when they are looked at in detail. For these there is no need to study or analyse their possible variability.

(A terminological comment: when in the following I mention the variability of a function, I really mean 'the variability of the Output from a function as a result of the variability of the function'. This is, however, too cumbersome. The two formulations 'the variability of a function' and 'the variability of the Output' are therefore used synonymously, unless when it is made clear that they are not.)

There can in principle be three different reasons for the variability of the Output from a function:

- The variability of the Output can be a result of the variability of the function itself, that is, due to the nature of the function. This can be considered as a kind of internal or endogenous variability.
- The variability of the Output can be due to the variability of the working environment, that is, the conditions under which the function is carried out. This can be considered as a kind of external or exogenous variability.
- The variability of the Output can finally be a result of influences from upstream functions, where the Outputs from upstream functions (as Input, Precondition, Resource, Control or Time) may be variable. This kind of coupling is the basis of functional resonance. It can also be called *functional upstream–downstream coupling*.

The variability of a function may, of course, also be due to combination (or rather, concatenation) of the three conditions, that is, internal variability, external variability and upstream–downstream coupling. In this section I will consider only the two first forms of variability. The third, variability due to functional upstream–downstream coupling, is the focus of Step Three of the method.

## Variability of Different Types of Functions

One way of characterising variability is to distinguish among different types of functions, for instance technological functions, human functions and organisational functions. (The choice of these three categories does not exclude other modes of description; but the choice is convenient and represents a widely used distinction.)

- Technological functions are carried out by various types of 'machinery'. The type of machinery depends on the working domain, whether it is an off-shore platform, an intensive care unit, a coal mine, an aircraft, a bank, an assembly line or something entirely different. In most cases the technology includes some form of computing machinery or information technology, which can collect, store, process, analyse and transmit data. This type of technology is often 'hidden' as part of the automation in the system. Since technological functions are designed to be highly predictable and reliable, the default assumption of the FRAM is that they do not vary significantly during the scenario that is analysed. This does not mean, of course, that technological functions cannot vary. Indeed, there are many examples of variability of technological functions, for instance as slow degradation due to wear and tear, as a consequence of the ambient conditions (for example, sensor malfunctions), because of software peculiarities, because of inadequate maintenance and so on. But unless there is reason to think otherwise, the default assumption of a FRAM analysis is that technological functions are stable.
- Human functions are carried out by humans, either as individuals (individual performance) or in small, informal (social) groups (collective performance). Chapter 3 has already discussed human performance variability in some detail, and listed the main influences or factors. In relation to safety, human performance has often been described as unreliable and 'error' prone, leading to the incorrect but nearly ineradicable notion that humans make errors. That is, however, judgemental and entirely misses the point that the variability of human performance is useful more often than it is harmful. Indeed, as argued throughout this book, performance variability is not only unavoidable but also indispensable for complex socio-technical systems. In a FRAM

analysis it is important to recognise that the frequency of human performance variability is high and that the amplitude is large. The high frequency means that performance can change rapidly, sometimes even from moment to moment. Humans respond very quickly to changes, not least in the interaction with others. The large amplitude means that differences in performance can be large, sometimes dramatically so – for better or for worse. The variations in both frequency and amplitude depend on many different things, including working conditions. One purpose of the FRAM is to produce a clear and comprehensive description of such dependencies.

- Organisational functions are carried out by a group or groups of people, sometimes very large groups, where the activities are explicitly organised. Organisations are obviously made up of people, but organisational functions differ from human functions by being described – and defined – on the level of the organisation as such. They are thus functions of the system per se, rather than of the people who work in the system. Organisations coordinate human functions to produce something that goes beyond what a disorganised group of individuals can do. For example, while an individual can teach, an organisation can provide training. And while an individual can check something, an organisation can provide quality control. For a FRAM analysis it is important to recognise that the frequency of organisational performance variability typically is low but that the amplitude is large. The low frequency means that organisational performance changes slowly, and that there usually also is high inertia. But the differences in performance, that is, the amplitude, can be large. The low frequency is easily exemplified by changes to rules, regulations or policies, which also illustrate the existence of inertia. For examples of large amplitude (or magnitude), consider the recurrent 'improvements' to security screening for air travellers.

This characterisation of performance variability has referred to what one may call the potential variability of a function, that is, whether it is likely to vary at all and, if so, how it may vary. The next step is to consider the potential variability in more detail, by looking at internal and external variability, respectively. The internal variability is an issue of how likely a function is to vary by itself, while the external variability is an issue of how likely a function is to vary as a consequence of the working conditions – which in turn may be seen as the outcome of other functions.

## Internal (Endogenous) Variability

In line with the above descriptions, I shall first consider possible internal reasons for variability in each of the three types of functions: technological, human and organisational. Later I shall consider the possible manifestations of the variability, that is, how it may express itself and how it may have effects for downstream functions.

For technology, performance can vary because the 'inner workings' of the technology are intractable; in other words, because we do not really know how the technology works, or sometimes not even how it is supposed to work. This can be the case for purely 'mechanical' systems, and is clearly the case for software systems. Variability can also be due to 'wear and tear', that is, the slow but inevitable degradation of mechanical components. But apart from that, there are no other major sources of internal variability for technology, though plenty of external ones.

For humans, functions can vary because of physiological and psychological factors. (Social factors will be considered an external source of variability. This is admittedly arguable since social factors may involve the perceived – or imagined – expectations of others.) Among the physiological factors, fatigue and stress (workload) are the ones that have been most thoroughly studied. Others are circadian rhythm, well-being (or illness), various physiological needs (relating to 'input' or to 'output'), temporary disabilities and so on.

As far as the psychological factors are concerned, this is a motley affair that includes such things as traits and biases, judgement heuristics, decision heuristics, problem-solving style, cognitive style and so on. To provide a complete, or even representative list, is out of scope for the current purpose, since there are far too many details. As mentioned above, social factors could be included here as a kind of psychological factor, if they describe what the person 'perceives' that others expect of him/her. For practical reasons I will nevertheless put them under external factors.

For organisations, performance can vary for several reasons, such as effectiveness of communication, authority gradient, trust, organisational memory, organisational culture (flexible, inflexible) and so on. The scientific literature offers several ways of characterising organisational culture, for instance, as a power culture, a role culture, a task culture and a person culture. For the discussion of a FRAM analysis the conclusion is that there are many ways in which organisational functions can vary in and of themselves.

This illustration of the many reasons for performance variability can support an initial assessment of whether performance variability is likely or not. The above considerations are summarised in Table 6.1.

**Table 6.1      Summary of internal (endogenous) variability**

|  | Possible internal sources of performance variability | Likelihood of performance variability (default assessment) |
|---|---|---|
| Technological functions | Few, well known | Low |
| Human functions | Very many, physiological and psychological | High frequency, large amplitude |
| Organisational functions | Many, function specific and/or relating to 'culture' | Low frequency, large amplitude. |

## External (Exogenous) Variability

In line with the above descriptions, I shall next look at possible external sources of variability for each of the three types of functions: technological, human and organisational.

The performance of technological functions can vary because of improper maintenance. This means that the technology or the 'machinery' does not meet the requirements or the design specifications. The origin is taken to be external, although one could argue that the need of maintenance is due to the internal wear and tear of the machinery in the first place. Other external factors are ambient operating conditions, particularly if they exceed the design specifications; sensor failures due to external conditions (for example, ice in Pitot tubes); overload, over-speed, excessive stress; improper use (meaning that the technology is used for a purpose different from the intended one) and so on.

For humans, the two main external sources are found in the technology and the organisation. Other sources are social factors, which can be other humans or the organisation, depending on how one looks at it. The social factors include things such as group pressures, implicit norms and so on, many of which overlap somewhat with organisational culture. From the organisation itself there are expectations, norms, demands, pressures, policies ('safety first', but...) and so on. The *Herald of Free Enterprise* is a good example of that, as the insistence on punctuality was important for how the vessel was declared ready for departure.

For organisations, the main influence is the operating environment, physical as well as legislative, and the business (commercial) environment. The influences include customer demands or expectations, availability of resources and parts, regulatory environment, commercial pressures, regulatory scrutiny (as in the

**Table 6.2     Summary of external (exogenous) variability**

|  | Possible external sources of performance variability | Likelihood of performance variability (default assessment) |
| --- | --- | --- |
| Technological functions | Maintenance, (mis)use | Low |
| Human functions | Very many, social and organisational | High frequency, large amplitude |
| Organisational functions | Many, instrumental or 'culture' | Low frequency, large amplitude |

financial crisis), as well as weather and ambient conditions (ash from erupting volcanoes or other whims of nature).

Even this brief discussion has shown that there are many reasons why functions can vary, and that no type of function is immune. Having said that, it is a default assumption of the FRAM that technological functions are relatively stable, that human functions may vary with high frequency and high amplitude, while organisational functions may vary with low frequency but high amplitude. This means that for any specific FRAM model, the variability of human and organisational performance is of most interest, regardless of whether it is the potential or the actual variability.

## Manifestations (Phenotypes) of Performance Variability

Having considered some of the possible sources of variability, internal as well as external, the next question is how performance variability will show itself – either in the sense of how it can be observed or detected – or in the sense of how it may affect downstream functions. There are two ways of doing that, either a simple (but robust) solution, which is efficient but not so thorough, or an elaborate solution which is more thorough but not as efficient. While the more thorough solution from an academic perspective should be preferred, the simpler solution is more practical. It can be used as a way of making a fast screening to identify those conditions where the more elaborate, but also more thorough, analysis may be warranted.

### *Characterising Performance Variability – The Simple Solution*

The simple solution to describe the consequences of performance variability is to note that the Output from a function can vary in terms of timing and precision. Strictly speaking, the FRAM does not consider the Output from a function as such but rather the Output as it is used by a downstream function – as Input, Precondition, Resource, Control or Time. The former would be akin to considering the quality of the Output in absolute terms, that is, the variability qua variability. In the spirit of the FRAM, it is more reasonable to consider the Output in its relation to a downstream function, that is, in a relative sense.

In terms of timing, an Output can occur too early, on time, too late or not at all. The last category, 'not at all', can be seen as an extreme version of 'too late'. It may mean either that the Output is never produced at all, or that it is produced so late that it, to all intents and purposes, is useless. It does, of course, play a role whether the Output was produced too late or simply reached the downstream function too late. If the latter is the case, there may have been a delay in the propagation. If so, the communication of the Output – or more generally, the transmission of the Output – should be considered as a function in its own right.

If one considers the typical characteristics of the various types of functions, referring once more to the common technology-human-organisation classification, the potential variability looks as shown in Table 6.3.

**Table 6.3      Possible Output variability with regard to time**

|  | Temporal range of variability of response | | | |
|---|---|---|---|---|
|  | Too early | On time | Too late | Not at all |
| Technological function | Unlikely | Normal, expected | Unlikely, but possible if software is involved | Very unlikely (only in case of complete breakdown) |
| Human function | Possible (snap answer), serendipity) | Possible, should be typical | Possible, more likely than too early | Possible, to a lesser degree |
| Organisational function | Unlikely | Likely | Possible | Possible |

In terms of precision, an Output can be precise, acceptable or imprecise. (More refined distinctions may, of course, also be used.) Since it refers to the coupling between upstream and downstream functions, the precision is relative rather than absolute. If the Output is precise, it satisfies the needs of the downstream function. A precise Output will therefore not in itself increase the variability of downstream functions, but may in fact possibly reduce it. An acceptable Output can be used by the downstream function, but the use requires some adjustment or variability of the downstream function, for instance that the information is interpreted or complemented as described by some of the ETTO rules. This may increase the variability of downstream functions. An imprecise Output is something that is incomplete, inaccurate, ambiguous or in other ways misleading. An imprecise output cannot be used as is, but requires disambiguation or interpretation, verification, a check against other information or against the situation as such – altogether something that may increase variability, typically by consuming resources and time that could and should have been used for other purposes. The consequences are of the same type as for an acceptable output, but their magnitude will be much larger.

**Table 6.4    Possible Output variability with regard to precision**

|  | Precision range of variability of Output | | |
|---|---|---|---|
|  | **Precise** | **Acceptable** | **Imprecise** |
| Technological function | Normal, expected | Unlikely | Unlikely |
| Human function | Possible, but unlikely | Typical | Possible, likely |
| Organisational function | Unlikely | Possible | Likely |

*Characterising Performance Variability – The Elaborate Solution*

The elaborate way to describe the consequences of performance variability uses the failure modes that are part of most safety models. These failure modes, or phenotypes, are the categories that describe how the variability of a function may be seen from the Output or the outcome of a function. Since functions and outcomes of necessity take place in a four-dimensional time-space continuum, it is possible to define an exhaustive set of failure modes as illustrated in Figure 6.1.

Rather than slavishly following the eight phenotypes and define eight different classification subgroups for the variability of the Output, it is practical to divide them into the following four subgroups (see Table 6.5):

- Variability in terms of timing and duration. This subgroup includes variability of the Output that is due to timing, speed and duration. For timing, the variability of the outcome is as

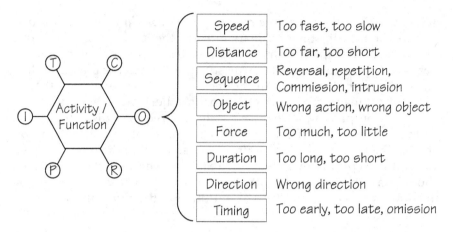

**Figure 6.1      Dimensions of failure modes**

described above. Duration means that an action can be too short (stopped before it was completed), or too long (continued even though it should have been stopped). Duration can be a consequence of doing something with the wrong speed, either too fast or too slow. In many cases there is a limit to how quickly something can be done; examples are braking with a car or cooling a material.

- Variability in terms of force, distance and direction. For force and distance, the Output can vary by being 'too little – too much' and 'too far – too short'. For direction, the possibilities are 'wrong direction' and 'wrong movement type'. Force denotes the power or effort that is produced by a function, for example, the level of thrust during take-off of an aircraft. If too much force is produced, equipment may malfunction or break; if insufficient power is produced, the action may not have the intended effect. Distance denotes the extension of the Output; it is synonymous with magnitude and may be linear or angular. Finally, direction denotes the line or course along which an Output moves.

- Variability in terms of the object that is the Output. In this case the Output refers to something that is moved, manipulated or otherwise affected by the function, for example, provided or displaced. This could, for instance, be a tool or a surgical instrument handed by the scrub nurse to the surgeon. The possible expressions of variability are that the object is the wrong object, either because it was a neighbour, a similar object or a completely unrelated object. Using the wrong object is one of the more frequent failure modes, such as pressing the wrong button, looking at the wrong indicator and so on.

**Table 6.5      Elaborate description of Output variability**

| Variability | Description/example |
|---|---|
| Timing / duration | Too early: A function is completed too early, executed faster than required or stopped before it should have been (premature output). |
| | Too late: A function is completed too late, executed with insufficient speed or continued beyond the point when it should have stopped (delayed output). |
| | Omission: A function does not complete at all (within the time interval allowed). |
| Force / Distance / Direction | Too weak or insufficient to accomplish desired effect. |
| | Too strong, excessive for the desired effect. |
| | A movement is too short or does not go far enough or is in the wrong direction. |
| | A movement is too long or goes too far or is of the wrong type. |
| Wrong object | The Output is the wrong object (for example, a tool) or points to the wrong object. The object can be a neighbouring object (proximity), a similar object or an unrelated object. |
| Sequence | Outputs that include a sequence of objects, of movements or of (state) changes, can vary in the following manner: Omission: a part of the sequence is missing. Jumping: a part of the sequence is skipped (either direction). Repetition: a part of the sequence is repeated when it should not have been. Reversal: two adjacent parts of the sequence change place. Wrong part: an extraneous or irrelevant part occurs in the sequence. |

- Variability in terms of the ordering or sequencing of the Output. Here the 'wrong place' refers to the sequence of the Output, as when the Output is made up of two or more steps or actions (or objects). This could for instance be the order of steps in an instruction, where a common mistake is to put the action before the test, such as in 'Stop Pump 23A when water level falls below 20 per cent' instead of 'When water level is below 20 per cent, then stop Pump 23A.'

*Potential and Actual Variability of Functions*

The FRAM model describes the possible couplings between functions, and is therefore also the basis for describing the potential variability of functions. Some examples of potential variability (without distinguishing between internal or external variability):

- For technology-based functions, the potential variability can be due to wear and tear, intractability (often referred to as complexity), inadequate maintenance, adverse operating conditions and so on. For example, if the temperature exceeds the design range for a piece of equipment, then it is possible that its performance will vary.
- For human based functions, the potential variability can be due to, for example, circadian rhythm, social pressure, decision-making habits and so on. For instance, if a function includes decisions it may potentially vary. Indeed, most if not all human functions are potentially variable. A classic example is the typical responses to input information overload. They describe what could possible happen when there is too much information, but not whether it will happen. That is only possible when further details about the people involved and the work situation become known.
- Finally, for organisation-based functions, the potential variability can be a consequence of limited memory, uncertainty about priorities (or conflicting priorities), ineffective communication (links) and so on.

Whereas the potential variability describes what could possibly happen if the circumstances were right (or wrong), the actual variability describes what should realistically be expected under given conditions, demands, opportunities and resources. Thus if the operating conditions (temperature and so on) for technology will be well within the design limits, then the potential variability is unlikely to become actual variability. Similarly, for human functions, if the activities take place in a shift-work setting, then it is reasonable to assume that there will be actual variability due to circadian rhythm. And finally, for organisational functions, if there are conflicting priorities, for example, between top management and the operating departments or between various groups of employees, then it is also reasonable to assume that the potential variability will be expressed as actual variability. The instantiation may be either an accident scenario, that is, something that has happened, or a potential future situation where risks or possibilities are considered. In the former case there is often quite detailed information, although it may not always be comprehensive. In the latter case the amount of detail depends on the assumptions that are made for the scenario.

It is important to distinguish between the potential and actual variability in the following step, namely how variability – or the consequences of variability – may spread through the system, specifically how the variability may affect downstream functions. The analysis of coupling and resonance should always refer to an instantiation of a model rather than the model, and therefore to the actual variability rather than the potential variability. (Actual variability does, of course not mean that it necessarily happens, but only that there is a significant likelihood that it will happen.) The actual variability will thus always be a subset of the potential variability. For this reason it is prudent to begin by describing the potential variability, in order to avoid being unnecessarily constrained by thinking of a specific scenario from the start.

# Chapter 7
# The Method: The Aggregation of Variability (Step 3)

The FRAM makes an important distinction between a model of the target system and the instantiations of the model. The FRAM model represents the set of functions that together account for the activity being analysed and the potential couplings among functions. An instantiation describes the couplings that existed or may exist for a given scenario or set of conditions, and thus represents a realisation of the model. (This has already been discussed in Chapter 6 with regard to assessing the potential versus the actual variability.) Since a FRAM model does not stand for any specific situation, it can only represent the potential performance variability. The range of this can be estimated based on established scientific knowledge as well as practical experience from a domain. An instantiation represents a concrete instance of the model for given (actual or assumed) circumstances and set(s) of conditions, and the details provided by the instantiation makes it possible to be more precise about whether and how the potential variability can become actual variability.

It is, however, not enough to know what the actual variability may be for individual functions. It is also necessary to know how variability may combine and thereby lead to outcomes that are either unexpected or out of scale – or both. In other words, it is necessary to know how functional resonance can come about. This is done by using the idea of functional upstream–downstream coupling.

## Functional Upstream–Downstream Coupling

As described in Chapter 6, the variability of a function can be due to the variability of the function itself or due to the conditions under which the function is carried out. The two were called internal and external variability, respectively. The variability of a function can finally be a result of couplings to upstream functions, where the Outputs from upstream functions (used as Input, Precondition, Resource, Control or Time) may be variable and thereby affect the variability of downstream functions. In order to describe the upstream–downstream couplings it is necessary to look in further detail at how differences in the quality of upstream Outputs can affect the variability of downstream functions, and thereby the variability of their Output. This will clearly depend both on the type of function, that is, whether it is a technological, a human or an organisational function, and on the specific nature of the function. In the case of the former, the general approach has already been described (Chapter 6). In the case of the latter, there are few short-cuts to suggest. Until a comprehensive taxonomy of the functions of a generic socio-technical system has been developed, each function must be described on its own using the combined knowledge of the analysing team.

## Upstream–Downstream Coupling for Preconditions

Preconditions are the system conditions or states that, in principle, must exist before a function can be carried out. Preconditions must obviously be provided by one or more other functions, that is, the state (of the Precondition) is the Output from upstream functions. In many cases, a function cannot be carried out unless the Preconditions have been established, and part of the function may be to check whether the Preconditions have been fulfilled, that is, whether the corresponding system state is true. One possibility is that the Preconditions are checked, but that the check cannot be completed because the timing of the 'signal' is imprecise, or because the state is ambiguous. The latter means that it cannot be determined precisely whether or not the Precondition is true. Another possibility is that the Preconditions are not checked properly due to variability in the way the checking is carried out.

If the state of a Precondition cannot be determined, this may increase the variability of downstream functions. It may, for instance, be necessary to wait until the Precondition has been established (loss of time), it may be necessary to query the upstream function (if possible) to verify or disambiguate information (again loss of time and/or resources), or it may be necessary to skip the check completely because of the need to remain synchronised with other activities. In either case, it is easy to imagine that the Output from the downstream function may vary in a characteristic manner, illustrated in Table 7.1.

(In the tables below, the possible change in variability is indicated as follows: [V↑] means that variability is likely to increase, [V↓] means that variability is likely to decrease (dampening), and [V↔] means that variability is likely to remain unchanged.)

**Table 7.1    Upstream–Downstream coupling for Preconditions**

| Upstream Output variability | | Possible effects on downstream function |
|---|---|---|
| Timing | Too early | False start; Precondition may be missed [V↑] |
| | On time | Possible damping [V↓] |
| | Too late | Possible loss of time [V↑] |
| | Omission | Increased improvisation, possible loss of time [V↑] |
| Precision | Imprecise | Possible loss of time (disambiguation); possible misunderstanding [V↑] |
| | Acceptable | No change [V↔] |
| | Precise | Possible damping [V↓] |

## Upstream–Downstream Coupling for Resources or Execution Conditions

Resources or Execution Conditions represent something that is needed and consumed by a function, or something that must be present while a function is carried out. If a Resource or Execution Condition is missing (omission of the Output from an upstream function), the function is blocked or hindered. If, for instance, the refuelling of an aircraft is late, the take-off will be delayed. If the refuelling cannot take place, the aircraft cannot fly at all. Likewise, if the correct tool (as an Execution Condition) is not available, then the function may take longer to carry out, and the efficiency or quality may possible also be reduced.

For Resources, it is clearly necessary that one or more functions either provide the Resources before the start or ensure that they are replenished whenever needed. For Execution Conditions, it is reasonable to assume that they are stable for the duration considered by the analysis. They can therefore be provided as the Output from background functions that are not described further for the scenario but which, of course, may become the focus of additional scrutiny. The functions that provide the Resources may, on the other hand, be either foreground functions or background functions, depending on how essential the Resources are or how much they are assumed to vary.

In some cases, the lack of a Resource or the absence of an Execution Condition may lead to a search for alternatives. This will at the very least mean that additional time is spent, hence introduce a delay in the Output from the function. If the alternative is not exactly what was needed (for example, in quantity or in functionality), then this may also impede the execution of the function and lead to inferior results – in other words lead to increased variability of the Output from the function. Conversely, fully adequate Resources or fully appropriate Execution Conditions – such as highly effective tools, of whatever kind they may be – may be able to dampen variability by permitting the function to be carried out a little faster or by ensuring that the Output is precisely as intended (and as required by a downstream function).

**Table 7.2    Upstream–Downstream coupling for Resources/Execution Conditions**

| Upstream Output variability | | Possible effects on downstream function |
|---|---|---|
| Timing | Too early | No effect [V↔] or possible damping [V↓] |
| | On time | Possible damping [V↓] |
| | Too late | Possible loss of time [V↑] |
| | Omission | Substitution by alternative approaches, if possible; improvisation [V↑] |
| Precision | Imprecise | Inadequate or reduced functioning [V↑] |
| | Acceptable | No effect [V↔] |
| | Precise | Possible damping [V↓] |

## Upstream–Downstream Coupling for Control

Control represents something that supervises or regulates how a function is carried out, and which therefore contributes to the quality of the Output – for instance in the sense that it is produced at the right time or in the sense that it is 'correct'. The Output from an upstream function that is the Control input for a downstream function can vary with respect to both time and precision.

Controls provide a further way to understand how functions may vary and how the couplings among functions can come about. The Control (input) to a function must clearly be the Output from one, or more, other functions. If the Control is the same for many situations, and if it is standardised such as an instruction or a procedure, then it is reasonable to treat it as a background function. In other words, it does not vary during the scenario. But if the Control is more active and adaptive, then it makes sense to treat it as a foreground function, which then of course has to be described in sufficient detail itself.

If the Control is imprecise, incomplete or even incorrect, then the function may vary as a result, that is, be carried out in a way different from what it ought to have been. But even a correct Control (input) may be used inappropriately, for instance if time is too short. Typical examples of this are described by the ETTO principle and specific ETTO rules; for instance, 'trust me, I have done this a million times before'. This in effect means that the person disregards the Control input for any number of reasons (habit, lack of time, social pressure and so on).

The temporal aspects of Control refer to whether the Control is available when it is needed or to the scheduling of the detailed steps or parts of a function. This is not least the case for instructions that are communicated or spoken, for instance when one operator reads the operating procedures aloud while another operator carries them out. Here an instruction may be issued prematurely, hence be missed, or issued too late. The situation could also be that the Control was not available or ready when it was needed (for instance because of retrieval problems, or because of variability of the function that provided it). Similarly, the information that comprises the Control, the 'instructions', can be imprecise or vague.

If the Control input is not available, the function may of course not be carried out, although it is more likely that the function will be carried out imprecisely or incorrectly (for example, 'I think

**Table 7.3    Upstream–Downstream coupling for Control**

| Upstream Output variability | | Possible effects on downstream function |
|---|---|---|
| Timing | Too early | Control input may be missed [V↑] |
| | On time | No effect, possible damping [V↓] |
| | Too late | Default or ad hoc control may be used instead [V↑] |
| | Omission | Substitute 'control' may be found [V↑] |
| Precision | Imprecise | Delays, trade-offs in precision and exactitude [V↑] |
| | Acceptable | No effect [V↔] |
| | Precise | Possible damping [V↓] |

it should be done in this way.') The same goes if the Control (the instruction) is imprecise. While it is always possible to use a call-back for confirmation (and in some domains it is even required), it is not always convenient either for practical reasons or because of working relations (for example, roving operators may try not to 'disturb' the central control room more than strictly necessary).

## Upstream–Downstream Coupling for Time

Time represents the various temporal relations that can have an influence on how a function is carried out. Time can, for instance, be the time that is available (the allowed duration), the (point in) time when a function can start or the (point in) time when it must be completed, the scheduling of a function relative to other functions, (synchronisation) and so on. One way of explaining the difference between Time and Control is to note that Time has to do with *when* functions are carried out while Control has to do with *how* functions are carried out.

Time also provides an important way to understand how functions can be coupled, and the definition of upstream and downstream functions obviously refers to time. The timing of functions can clearly have an influence on the variability of performance and therefore also on the variability of the Output. Too little Time, whether it is because a function is started too late, because it must be completed too early, or because it goes slower than expected or usual, will typically lead to trade-offs or sacrifices in the way it is carried out. One consequence of that may be that some Preconditions are skipped,

Table 7.4    Upstream–Downstream coupling for Time

| Upstream Output variability | | Possible effects on downstream function |
|---|---|---|
| Timing | Too early | Jump start, incorrect timing [V↑] |
| | On time | No effect, possible damping [V↓] |
| | Too late | Delayed activity; scheduling conflicts; short-cuts in performance; loss of synchronisation [V↑] |
| | Omission | Imprecise or incorrect start or stop of function [V↑] |
| Precision | Imprecise | Increased variability [V↑] |
| | Acceptable | No effect [V↔] |
| | Precise | Possible damping [V↓] |

on the assumption that they – as 'usual' – will be true, or that some Controls are not followed in detail. (This is one of the main arguments of the efficiency-thoroughness trade-off principle.) This is so whether it is the occurrence or the contents of the Time input that is imprecise.

Many individual (and collective) human functions depend on Time, either as a (synchronisation) signal or as a duration. Organisational functions are typically constrained mainly by duration. This is often because other (organisational) functions may depend on them, for example, such as the completion of a project, a clearance or an investigation. Organisational functions are also typically of longer duration than individual human functions.

## Upstream–Downstream Coupling for Input

The Input, finally, represents that which starts or initiates a function, as well as that which is used or transformed by the function. As described above, all the other aspects – and in particular Time and Control – can also be seen as special kinds of inputs.

The role of the Input as a signal that marks the beginning of a function is useful to understand how the variability of functions can arise. For instance, if the function is 'looking' for a signal in order to begin, then the performance can vary if the function 'looks' at the wrong signal, if the detection threshold is either too low or too high, if the Input is misinterpreted (leading either to a lack of response even though the signal was present, or a response even though the signal was not present) and so on. The lack of uniformity in starting the function can be due either to variability of the Input itself (which provides a coupling to the Output from an upstream function) or to the variability of the function that receives the Input.

If the Input is used to start a function, then the variability of timing is clearly important. And here it obviously makes a big difference if a function begins too early or if it begins too late. In the first case there may be problems of synchronisation that lead to delays with effects that spread through the instantiation. In the second case, which probably is more frequent, synchronisation problems may lead to a compression in how the function is carried out, typically in the form of one or several trade-offs or sacrifices. Something has to be done to gain time (or to conserve resources), and this increases the variability of the function. If enough time is gained, the consequences for downstream functions may – ironically – be reduced, since the faster execution in a way absorbs the delay. But the precision of the Output may suffer from that.

If the Input is taken in or operated on by the function, then variability of precision is more important than variability of timing. Data or information may be imprecise, equivocal or noisy and matter and materials may be irregular, impure or contaminated. If the Input is imprecise, then more time may be needed to process it, hence introduce delays. The processing may also lead to the wrong results, since the Input may be misunderstood or misinterpreted.

Table 7.5     Upstream–Downstream coupling for Input

| Upstream Output variability | | Possible effects on downstream function |
|---|---|---|
| Timing | Too early | Premature start; Input possibly missed [V↑] <br> No effect, possible damping [V↓]* |
| | On time | No effect, possible damping [V↓] |
| | Too late | Function delayed, leading to short-cuts [V↑] |
| | Omission | Function not carried out or severely delayed [V↑] |
| Precision | Imprecise | Loss of time, loss of accuracy, misunderstandings [V↑] |
| | Acceptable | No effect [V↔] |
| | Precise | Possible dampening [V↓] |

* In cases where the downstream function is just waiting for the input, an early input may provide additional time, hence possibly dampen variability.

## Issues in Upstream–Downstream Coupling

The discussion of upstream–downstream couplings has tacitly assumed that it always is possible to decide whether the Output from an upstream function is the Input, Precondition, Resource, Control or Time to the downstream function. But this assignment may in some cases be open to dispute. In such cases, the analysts should consider the primary use or role of an Output, and let that determine whether it is, for example, a Time or a Control aspect. In the long run, such uncertainty need not be fatal since the functional coupling itself is more important than whether it is via one aspect or another.

The upstream–downstream couplings can describe how the variability of functions may spread in a way that is fundamentally different from the usual linear propagation. One reason is that the 'path' is not predefined but depends on how an instantiation develops. The FRAM model defines the potential couplings, but the instantiation shows which of these become actual under the given conditions. The correctness of this obviously depends upon how carefully the model has been developed, how precisely the functions have been characterised, and how realistic the instantiations are. This proviso, however, applies to any analysis method. But given that, the FRAM can describe how the couplings may combine so that everyday performance variability rather than failures and malfunctions can lead to unexpected outcomes, and also how out-of-scale or non-linear outcomes can arise. This is the functional resonance that has given the method its name. At the moment the outcomes have to be derived in a step-by-step fashion, and the approach is qualitative rather than quantitative. The simplicity and the consistency of the method, however, means that it should be straightforward to build supporting software, and probably also to find a way of calculating the magnitude of the functional resonance.

## Comments on Chapter 7

The ETTO principle, also briefly mentioned in Chapter 2, is the idea that all human activity – individual, collective or as an organisation – can be described as if it involved a trade-off between efficiency and thoroughness. This trade-off is sometimes deliberate but is in most cases habitual and almost automatic, hence goes unnoticed. The efficiency-thoroughness trade-off is one way of accounting for the ubiquitous performance adjustments. The principle has been described in Hollnagel, E. (2009), *The ETTO Principle: Efficiency-Thoroughness Trade-Off. Why Things that Go Right Sometimes Go Wrong*, Farnham, UK: Ashgate.

# Chapter 8
# The Method: Consequences of the Analysis (Step 4)

The fourth and final step in the FRAM is to propose ways to manage the possible occurrences of uncontrolled performance variability – or possible conditions of functional resonance – that have been found by the preceding steps. If uncontrolled performance variability may lead to adverse and unwanted outcomes, the purpose is of course to make sure this does not happen. But since the FRAM is about performance variability regardless of whether it is positive or negative, it is also possible that uncontrolled performance variability may lead to positive or desirable outcomes. In such cases, the purpose is to facilitate and possibly enhance the outcomes without losing control. Having said that, the rest of this section will nevertheless focus on the possibility of adverse outcomes, mainly because this has been the traditional focus in safety studies. (Resilience engineering, of course, advocates that one looks at positive outcomes as well; see the ETTO principle.)

As a starting point for discussing the contents of Step 4, it is useful briefly to look at how the basic thinking about the nature of accidents, the fundamental assumptions that people make, affects what is done (see Table 8.1), both in the sense of what an investigation typically looks for, and what the ensuing actions or responses are.

If the thinking about how accidents occur – the accident model – explains them in terms of one or more sequences of causes and effects, that is, by a simple sequential model, then the purpose of the investigation is to find specific causes such as components or functions that somehow failed, and to identify the cause–effect links that can explain the outcomes. (This is, for instance, the basic principle of the root cause analysis, and also characterises event trees and fault trees.) The focus of the recommendations is, not surprisingly, to prevent accidents from happening either by eliminating potential causes (hazards) or by disabling cause–effect links. In addition, steps may also be proposed to provide protection against the outcomes that we know will occur despite all precautions.

**Table 8.1    Three accident philosophies**

|  | Basic principles | Purpose of investigation | Focus of recommendations |
|---|---|---|---|
| Simple, linear model (for example, Domino) | Causality (single or multiple causes) | Find specific causes and cause–effect links | Eliminate causes and cut cause–effect links |
| Complex linear model (for example, Swiss cheese) | Latent conditions, hidden dependencies | Find combinations of unsafe acts and latent conditions | Strengthen barriers and defences |
| Non-linear (systemic) model | Dynamic couplings, functional resonance | Find tight couplings and complex interactions | Monitor and manage performance variability |

If the thinking about how accidents occur explains them in terms of a combination of ongoing events and previously existing conditions, that is, by a complex sequential or epidemiological accident model, then the purpose of the investigation is to find the critical combinations of unsafe acts and latent conditions, where the latter often are expressed as weakened defences or failed barriers. The focus of the recommendations is consequently on how barriers and defences can be strengthened, but also on how latent conditions can be detected – and neutralised – before they possibly contribute to an accident.

Finally, if the thinking about how accidents occur assumes that they are best understood in terms of dynamic couplings that go beyond linear cause–effect reasoning, that is, as emergent and non-linear outcomes of dynamic system processes, then the purpose of the investigation is to find the tight couplings and complex interactions that can explain the outcomes. The focus of the recommendations also changes to look at how performance variability can be monitored and managed, specifically how it is possible to dampen variability without at the same time rendering the system dysfunctional.

The three basic accident models described in Table 8.1 are of course not mutually exclusive, but rather differ in terms of the situations for which they are effective. The simple linear models can be applied to the large number of accidents and incidents where the event justifiably can be explained by a few factors and where the activities are fairly regular and homogeneous. But simple linear models are less well suited to events where the outcome is due to the interaction of many factors, where some may be active in the event and others be latent. In such cases the complex linear models are more appropriate, particularly if they also address organisational and cultural issues. Yet the growing number of events where the outcomes are emergent rather than resultant defy even complex linear models. In such cases it becomes necessary to focus on functions rather than structures, and to develop ways to represent and understand the dynamics of complex socio-technical systems. The FRAM represents such a non-linear approach.

When an accident is explained in terms of non-linear dynamic dependencies among functions rather than simple or complex cause–effect relations, the consequences of the explanation are different. The dilemma facing experts and safety managers is that the situation cannot be improved by *eliminating* performance variability, since performance variability is essential to ensure both safety, productivity and quality. The solution is instead to *manage* performance variability, typically by trying to dampen the variability in order to weaken resonance effects.

## Elimination, Prevention, Protection and Facilitation

The primary purpose of the FRAM is to build a model of the functions of a system that describes how performance variability may occur in everyday operations and how the effects may spread through the system. This provides the necessary basis for identifying potential problem areas in the system's functioning, in addition to the more traditional analyses of failure modes, malfunctions and so on. Once such problem areas have been found, a number of well-known remedies can be applied:

- The classical safety solution is to *eliminate* the hazards that are found by removing the offending parts or components of the system. This can either be a physical part or a certain operation or activity. Although elimination is effective, it is a relatively rare occurrence, since it may require a complete redesign of the system or the activity. If elimination only is done partially, it falls prey to the vagaries of the substitution myth.
- Another commonly used solution is *prevention*, which typically takes place by introducing some kind of barrier or defence. Barriers can be of several types, where a common classification distinguishes between physical, functional, symbolic and incorporeal barriers. The choice of barrier system invariably involves a trade-off between speed (efficiency) and effect (thoroughness). Complex and resource demanding barriers, typically physical and functional barriers, are costly and take

time to put into place, but are also very effective. Simpler barriers, often of the symbolic type, may be both fast and inexpensive to set up, but the effect may soon wear off and thereby render the barrier system ineffective.

- Whereas prevention is aimed at that which is harmful, *facilitation* is aimed at that which is useful. This could involve changing or redesigning the system so that it is easier to use correctly – and more difficult to use incorrectly. Facilitation can be used to make tasks less complex, for instance through task redesign, improvements to the human-system interface, or by providing various types of operational support for users. Facilitation is congruent with the resilience engineering interpretation of safety as the ability to do things right.
- *Protection* is focused at the outcomes from an accident rather than the events that lead to them. It can involve the use of various types of barriers to contain the consequences or the provision of means of mitigation or of recovery. Protection is by its very nature reactive but paradoxically requires considerable imagination when it is put in place.

There is a fairly extensive literature, including guidelines and practical examples, to help with the choice and implementation of these solutions. Many of them have been used for decades, and different industrial domains often have strong traditions for how risks are treated, which reflect both how mature the technology is and how often serious accidents happen. Some or all of these solutions can, of course, be used for weaknesses that are identified using the FRAM. But the FRAM also proposes two additional solutions, namely monitoring and dampening.

## Monitoring (Performance Indicators)

Control in a socio-technical system, such as safety or quality management, requires that appropriate *targets can be set*, that ongoing *processes* and *developments can be monitored* and that *activities can be regulated* to keep the processes on track so that both short-term and long-term goals can be attained.

The selection of which indicators to monitor often involves a compromise among several criteria and interests – a trade-off between efficiency and thoroughness. While the literature offers extensive lists of criteria for indicators, there are few discussions of the reasons behind their selection. And although it clearly is more important that indicators are meaningful than easy to use, the choice of an indicator usually favours the latter rather than the former. Meaningfulness requires a good understanding of what goes on in a system. For technological systems this can be based on the knowledge of the design and architecture (the parts, their qualities, their interrelations, their basic behaviour and so on). For socio-technical systems there is no comparable basis, and the understanding of what goes on in the system is often incomplete, if not inaccurate (see Chapter 1). One reason is that system models traditionally represent the flow of something (data, matter, energy and so on) between components or parts, hence refer to system structures rather than system functions.

A FRAM model describes the functions of a system and how they are mutually coupled. A FRAM model can in particular be used to identify the conditions where developments potentially may get out of control, for instance by identifying couplings that may lead to an increase in performance variability. A FRAM model can therefore be used as a basis for proposing indicators, hence as a basis for monitoring. Indeed, since a precondition for the ability to manage a system or a set of processes is that the current state of the system is known, some kind of functional model is indispensable. The value of doing so can be demonstrated by developing a FRAM model of general process management, although that is beyond the scope of this book.

## Dampening

The effective management of a system or a set of processes not only requires that the current state is known, but also that the means for effective intervention in the process are available. The means should enable those in charge of the system to make sure that developments go in the right direction and progress at the right speed.

As summarised by Table 8.1, the recommendations based on non-linear or systemic thinking focus on ways to monitor and manage performance variability, in line with the four principles underlying the FRAM (Chapter 3). In the language of process control this means that the variability must be managed, specifically that it must be dampened when it looks like it will get out of hand and lead to adverse outcomes.

The previous chapters have described how performance variability can be due to internal characteristics (endogenous), external conditions (exogenous) and upstream–downstream couplings. For the purpose of managing performance variability, the options are limited for variability that is due to internal characteristics. For humans it may possibly be an issue of selection, while for organisations it may be an issue of inherently superior organisational design, if that really exists.

For variability that is due to external conditions, specifically the conditions under which the work is carried out, solutions are possible but usually not very practical in a concrete situation. The reason is simply that it may take time – and resources – to change the working conditions, hence to achieve the desired effects on the variability. Since such changes furthermore will be of a permanent rather than a temporary nature, they should not be introduced unless it is certain (with relative and acceptable ignorance, of course), that there will be no unintended consequences.

This leaves the variability that is due to upstream–downstream couplings. This is also the variability that likely will be most volatile, as it is subject to the current – and constantly changing – conditions. The discussion of the different types of upstream–downstream couplings in Chapter 7 described both the conditions under which performance variability could increase and the conditions under which it could be dampened. The conclusions from this discussion are summarised in Table 8.2. Not surprisingly, variability can be dampened by reducing the variability in the Output from upstream functions.

**Table 8.2    Consequences of upstream variability on downstream variability**

| Upstream Output variability | | (I) | (P) | (R) | (C) | (T) |
|---|---|---|---|---|---|---|
| Timing | Too early | [V] [V⁻]* | [V] | [V«], [V⁻] | [V] | [V] |
| | On time | [V⁻] | [V⁻] | [V⁻] | [V⁻] | [V⁻] |
| | Too late | [V] | [V] | [V] | [V] | [V] |
| | Omission | [V] | [V] | [V] | [V] | [V] |
| Precision | Imprecise | [V] | [V] | [V] | [V] | [V] |
| | Acceptable | [V«] | [V«] | [V«] | [V«] | [V«] |
| | Precise | [V⁻] | [V⁻] | [V⁻] | [V⁻] | [V⁻] |

* In cases where the downstream function is just waiting for the input, an early input may provide additional time, hence possibly dampen variability.

Since this variability in turn depends on other functional couplings rather than on internal or external sources, it means that there are no short-cuts or easy solutions. In particular, there are no root causes to be addressed. While this may appear to make the safety manager's life more difficult, the silver lining is that the dampening not only serves the purpose of preventing things from going wrong, but also contributes to things going right. In other words, the purposes of safety and productivity become synergistic rather than antagonistic – in good agreement with resilience engineering thinking.

## What About Quantification?

Quantification has always played a significant role in risk assessments. This is clear from the very name of the commonly used approaches, namely Probabilistic Risk Assessment (PRA) and Probabilistic Safety Assessment (PSA). A probability is a numerical indication of the likelihood that something happens, ranging from a value of 1.0 for something that is absolutely certain to a value of 0.0 for something that is absolutely impossible. In the context of safety, probability values are usually low, for instance $10^{-5}$ or $10^{-6}$, meaning that the likelihood of something going wrong is 1/100,000 or 1/1,000,000. A rather spectacular example of this way of thinking can be found in the project for sending the first spacecraft to Mars in the late 1970s. Since it was possible that a spacecraft from Earth might contaminate Mars, it was stipulated that the probability of this happening should be less than one billionth ($p < 10^{-9}$).

While it may be reasonable to express the outcome of a technical safety analysis, a PSA, using probabilities, it is inappropriate when it comes to analyses including human performance. It has nevertheless become a de facto practice to try to calculate the probability of a failure by human operators, (in)famously known as a 'human error probability' or HEP. The need to do so was a consequence of transferring the technical analysis methods to cover also human performance – and later organisational performance as well. The set of methods developed to provide these outcomes, to provide the quantification, are commonly known as human reliability assessment (HRA) methods.

Because of this need, and the associated established practice, there will inevitably be a question of whether the outcomes of a FRAM analysis can be expressed as probabilities, at least partly. In other words, can quantification become part of the FRAM? Before trying to give an answer to that, it is reasonably to consider whether the question is meaningful, which is another way of asking whether quantification is necessary?

In the case of the FRAM, the focus is on variability rather than probability. So the question becomes whether it is meaningful to express variability quantitatively. There is no established tradition for doing so in PRA or HRA, which at best supply probabilities with uncertainty intervals (lower and upper bounds). An uncertainty interval is, however, an expression of range rather than of variability. At the current stage of development of the FRAM, the variability is characterised verbally, that is, as a quality rather than a quantity. It is entirely possible to introduce fuzzy rather than crisp descriptors, and this may be a step in the direction toward quantification. It is important to keep in mind, however, that the purpose of the FRAM is to represent the dynamics of the system rather than to calculate failure probabilities. In fact, the concept of failure is not part of the FRAM at all. And whether it is useful to work with quantitative rather than qualitative descriptions of variability remains to be seen. The conclusion is therefore that the question about quantification may not really be meaningful.

## Comments on Chapter 8

The importance of accident models for how investigations are conducted can be seen from Lundberg, J., Rollenhagen, C. and Hollnagel, E. (2009), 'What-You-Look-For-Is-What-You-Find – The consequences of underlying accident models in eight accident investigation manuals', *Safety Science*, 47, 1297–1311.

The substitution myth is the belief that artefacts are value neutral in the sense that they will only have the intended but no unintended effects when they are introduced in a system. The facts are that any change to a system, not least a partial elimination of a hazard, will have consequences that often are unexpected. The substitution myth was first described in the field of human–computer interaction by Carroll, J.M. and Campbell, R.L. (1988), *Artifacts as Psychological Theories: The Case of Human–Computer Interaction*. User Interface Institute, IBM T.J. Watson Research Centre, Yorktown Heights, New York.

A comprehensive treatment of barriers, including the explanation of the various barrier systems, can be found in Hollnagel, E. (2004), *Barriers and Accident Prevention*, Aldershot, UK: Ashgate. (This book also provides the first description of the FRAM.)

Fuzzy descriptions are used to describe sets where the members can have various degrees of membership. It basically means that terms, such as 'a high likelihood', can be used in a consistent manner so that it is possible to draw conclusions (using fuzzy logic) from descriptions that seem to be vague or imprecise. Fuzzy set theory was proposed by Zadeh, L.A. (1965), 'Fuzzy sets', *Information and Control*, 8(3), 338–353.

# Chapter 9
# Three Cases

This chapter presents three short cases to show how the FRAM can be used in practice to model how a system works as a basis for understanding an event. Two of the three cases are events that have happened, one being a surgical incident and the second a serious maritime accident. The third case is a risk assessment, that is, an analysis of a system to find possible situations or conditions where things may go wrong. For none of the three cases is a complete model provided, simply because that would be too long for a book chapter. All three cases have been described in the open literature, and for the surgical case a FRAM analysis has also been presented. Details about this can be found in the comments to this chapter.

## 'Duk i buk' (The Sponge in the Abdomen)

This case describes a surgical incident where material unintentionally and regrettably was left in a patient's abdomen. This was discovered after the wound had been sutured. The patient had been extubated, but had not left the operating room. The patient immediately underwent a new – and successful – operation in order to retrieve the material.

### The Event as It Happened

The surgical procedure was performed by two surgeons working as a team. Prior to the operation they had agreed on their roles as main and assisting surgeon, respectively. The main surgeon was a specialist who was very familiar with the procedure. The assisting surgeon was also a specialist. They had both been asked whether they could perform two urgent surgical procedures on patients in other operating theatres during the same time period, and they had both agreed to do so. This was something that happened occasionally in the hospital. In such cases the surgeons usually tried to perform any additional surgical procedure in between the planned ones.

In the present case, the main surgeon left the Operating Room (OR) towards the end after suturing the wound, in order to perform surgery in another OR. The assisting surgeon left the OR twice during the event to perform urgent surgery in other ORs – once after the tissue sample had been excised and once after the haemorrhage had been stopped. One consequence of the simultaneous operations was that the telephone in the main OR rang frequently during the surgery, because staffs in other ORs were waiting for the surgeons to arrive. This was, not surprisingly, seen as stressful.

The designated main surgeon initiated the abdominal surgical procedure together with the designated assisting surgeon. When the tissue sample had been excised, the assisting surgeon left the OR for the first time to begin working on one of the other urgent procedures. Measures to stop the bleeding (haemostasis) were at the same time undertaken for the surgical procedure in the main OR.

When the assisting surgeon returned, the main surgeon asked him to come to the wound to assist with the haemostasis, which posed a problem. Large amounts of sponges were consequently consumed. The scrub nurse counted all the instruments and materials when an abdominal sponge (a large cloth for haemostasis) and a disarp (a disposable abdominal retracting pad) were still in the patient's abdomen but did not report the result to the surgeon. Since the patient was taking part in a study, the scrub nurse prepared a syringe with a special analgesic that was to be used in the wound. The assisting surgeon removed the retractor and asked the main surgeon to close the wound herself, without his assistance. The assisting surgeon then left the OR for the second time to perform another surgical procedure elsewhere.

At this time the main surgeon was informed that her next patient was ready for the pre-operative briefing. She informed the staff waiting for her that she would arrive as soon as she was done with the suturing (closing the wound). She started suturing the patient in the main OR, and asked the scrub nurse to assist. Neither the scrub nurse nor the supervising nurse made an extra check of materials or reported that materials might have been left in the abdomen.

Concurrently, the supervising nurse, who then primarily performed the tasks assigned to the assistant nurse, checked the papers regarding the study and realised that the syringe had been prepared with the wrong analgesic. The three nurses (the scrub nurse, the supervising nurse and the assistant nurse) had a discussion to determine which analgesic to use. They discarded the syringe, and the scrub nurse prepared a new one with the right analgesic. During this the telephone rang again as the staff of the main surgeon's next surgical procedure anew wondered when she would join them. The main surgeon left the OR after the analgesic had been given and the wound had been sutured.

The scrub nurse and the supervising nurse did a final count of the instruments and materials and realised that an abdominal sponge and a disarp were missing. Everyone in the OR was informed and attempts were made to contact the two surgeons both by telephone and pager. The nurse anaesthetist was just waking the patient, who had already been extubated, but the patient was quickly intubated and anaesthetised again. In the situation, the nurse anaesthetist would have liked to have support from the anaesthetist to intubate and re-anaesthetise the patient. This was unfortunately not possible because the only telephone in the OR was occupied. The supervising nurse eventually got hold of the main surgeon who confirmed that she would return immediately to retrieve the material.

A new surgical procedure was quickly prepared, the abdomen was opened and the materials removed. The patient did not wake up sufficiently to remember anything of what had happened, but was informed about it afterwards. Fortunately, the patient did not suffer any permanent harm.

## The Functions and the FRAM Model

Following the principles of the FRAM, the first step is to identify the functions that are required for everyday work in the case. A relatively simple, and non-technical, description is as follows:

- Prepare the patient for surgery (anaesthetise the patient).
- Begin the surgical procedure.
- Perform the surgical procedure (extract tissue sample).
- Complete the surgical procedure (apply special analgesic; suture the wound).
- Wake the patient.

Since it is not practical to go through a complete FRAM analysis here, the presentation will focus on the functions that should have been in place for the completion of the surgical procedure to be carried out successfully. This makes sense, since it was here that performance varied so much, that it led to an unwanted outcome.

As the description of the event makes clear, several people were involved in the event and what one person did (or did not do) affected what other persons did. It is therefore reasonable to focus on the overall work, rather than on what specific persons did. The event involved the two surgeons and the three nurses who have already been mentioned. The scrub nurse had recently passed the examination and took part in an orientation programme at the time. The assistant nurse was only present during the preparations and the final part of the surgery. In addition there were two other people, a nurse anaesthetist and an anaesthetist. The latter was not present in the OR during the surgical procedure but was ready nearby in case the nurse anaesthetist needed help.

This surgical procedure differed from usual practice because the patient took part in a study. This required a special analgesic to be administered towards the end, after the surgical procedure had been completed but before the wound was sutured.

With these conditions in mind, we can begin to identify and describe the functions of interest. To make the description as brief as possible, the first pass through the analysis only describes the aspects that are of immediate concern for the analysis. A second pass may both elaborate on those descriptions and complement them with descriptions of other aspects. A good starting point is the function called <Suturing the wound>.

**Table 9.1        FRAM representation of <Suturing the wound>**

| Name of function | Suturing the wound |
|---|---|
| Description | This is part of the completion of the surgical procedure. To be done by the main surgeon |
| **Aspect** | **Description of aspect** |
| Input | Surgical procedure has been completed |
| Output | Wound has been sutured |
| Precondition | All instruments and materials are accounted for |
|  | Special analgesic has been administered |
| Resource | *Not described initially* |
| Control | *Not described initially* |
| Time | *Not described initially* |

The description of the aspects of this – first – function points to four other functions. One is the downstream function that follows the completion of the surgical procedure, that is, the function that has [Wound has been sutured] as Input. The second is the upstream function that as Output has [Surgical procedure is complete]. And the third and fourth are the two upstream functions that as Output have the two Preconditions described for <Suturing the wound>, namely [All instruments and materials are accounted for] and [Special analgesic has been administered]. Note that the function as described here has two Preconditions. In general, a function can have as many instances of an aspect as is required. That goes for all the six aspects.

The FRAM analysis is continued by describing each of the four functions identified above. Since they all have to be described, the order in which it is done does not matter.

**Table 9.2     FRAM representation of <Count instruments and materials/AS>**

| Name of function | Count instruments and materials/AS |
|---|---|
| Description | To be done by the scrub nurse after suturing is completed |
| **Aspect** | **Description of aspect** |
| Input | Wound has been sutured |
| Output | Patient is ready for waking |
| Precondition | *Not described initially* |
| Resource | *Not described initially* |
| Control | *Not described initially* |
| Time | *Not described initially* |

The counting of instruments and materials should have been done several times during the surgery, at least both when the surgical procedure had been completed but before the wound was sutured and again after the wound was sutured but before waking the patient. In order to distinguish between the two different counting functions they are marked as BS (Before suturing) and AS (After Suturing), respectively.

The Output is that the patient is ready for waking, because all instruments and materials have been accounted for. The counting is carried out after the wound has been sutured, as a final check before extubating and waking the patient. It points to two other functions, one of which has already been described. The other follows below.

**Table 9.3     FRAM representation of <Waking the patient>**

| Name of function | Waking the patient |
|---|---|
| Description | This is done by the nurse anaesthetist, possibly assisted by the anaesthetist, after the surgical procedure has been completed |
| **Aspect** | **Description of aspect** |
| Input | Patient is ready for waking |
| Output | *Not described initially* |
| Precondition | *Not described initially* |
| Resource | *Not described initially* |
| Control | *Not described initially* |
| Time | *Not described initially* |

In the FRAM analysis, the function <Waking the patient> is in the first iteration considered as a background function. The function is required as a destination for the Output [Patient is ready for waking]. Background functions can be used to ensure that all described aspects of a function have potential couplings to other functions. Since <Waking the patient> in this case simply serves as a sink for the Output from an upstream function, it does not need to have an Output itself. Because none of the aspects are described, <Waking the patient> does not point to any additional functions. It therefore constitutes a natural ending point of the analysis. For the same reason, it also defines the boundary of the system being analysed. The <Waking the patient> is in practice a placeholder for a function, which may need to be described in further detail later depending on how the analysis turns out. This provides the way by which a FRAM model can be extended or elaborated upon.

As noted above, the instruments and materials were counted on several occasions.

Table 9.4    FRAM    representation    of    <Count    instruments    and materials/BS>

| Name of function | Count instruments and materials/BS |
|---|---|
| Description | To be done by the scrub nurse before suturing starts |
| **Aspect** | **Description of aspect** |
| Input | Surgical procedure has been completed |
| Output | All instruments and materials are accounted for |
| Precondition | Special analgesic has been administered |
| Resource | *Not described initially* |
| Control | *Not described initially* |
| Time | *Not described initially* |

Since the patient took part in a study, a Precondition for the function <Count instruments and materials/BS> was that [Special analgesic had been administered]. This function therefore points to two other functions, of which one has already been identified by <Suturing the wound>. Since the performance of <Count instruments and materials> was variable, it is considered to be a foreground function rather than a background function.

The administration of the special analgesic was not part of the common surgical procedure. It was, however, part of this event, and is therefore included among the functions that should be carried out.

It has also been pointed to by <Suturing the wound> and <Count instruments and materials>.

**Table 9.5    FRAM representation of <Administer special analgesic>**

| Name of function | Administer special analgesic |
|---|---|
| Description | To be done by the scrub nurse. Note that this was an additional task during this surgery |
| **Aspect** | **Description of aspect** |
| Input | Tissue sample has been excised |
| | Syringe with special analgesic |
| Output | Special analgesic has been administered |
| Precondition | Haemostasis has been achieved |
| Resource | *Not described initially* |
| Control | *Not described initially* |
| Time | *Not described initially* |

The function <Administer special analgesic> points to three other functions, which need to be briefly described. For the event in question they are all foreground rather than background functions, one reason being that they could potentially vary during the time frame considered. They therefore require at least an Input defined. But in all cases the Input is assumed to be the Output of a background function, which means that the analysis need not go any further. The <Prepare special analgesic> can be described as follows:

**Table 9.6    FRAM representation of <Prepare special analgesic>**

| Name of function | Prepare special analgesic |
|---|---|
| Description | To be done by the scrub nurse |
| **Aspect** | **Description of aspect** |
| Input | Abdominal surgery has been initiated |
| Output | Syringe with special analgesic |
| Precondition | *Not described initially* |
| Resource | *Not described initially* |
| Control | *Not described initially* |
| Time | *Not described initially* |

The <Excise tissue sample> can be described as follows:

**Table 9.7      FRAM representation of <Excise tissue sample>**

| Name of function | Excise tissue sample |
|---|---|
| Description | To be done by the main surgeon |
| **Aspect** | **Description of aspect** |
| Input | Abdominal surgery has been initiated |
| Output | Tissue sample has been excised |
| Precondition | *Not described initially* |
| Resource | *Not described initially* |
| Control | *Not described initially* |
| Time | *Not described initially* |

The required background function is in both cases <Initiate abdominal surgery>.

**Table 9.8      FRAM representation of <Initiate abdominal surgery>**

| Name of function | Initiate abdominal surgery |
|---|---|
| Description | To be done by the main surgeon |
| **Aspect** | **Description of aspect** |
| Input | *Not described initially* |
| Output | Abdominal surgery has been initiated |
| Precondition | *Not described initially* |
| Resource | *Not described initially* |
| Control | *Not described initially* |
| Time | *Not described initially* |

Since this is considered to be a background function only the Output is defined. None of the five other aspects is described any further.

The <Achieve haemostasis> can be described as follows:

**Table 9.9    FRAM representation of <Achieve haemostasis>**

| Name of function | Achieve haemostasis (stop bleeding) |
|---|---|
| Description | This describes the steps needed to stop haemorrhage (bleeding) |
| **Aspect** | **Description of aspect** |
| Input | Uncontrolled bleeding |
| Output | Haemostasis has been achieved |
| Precondition | *Not described initially* |
| Resource | *Not described initially* |
| Control | *Not described initially* |
| Time | *Not described initially* |

The required background function is <Haemorrhage>. This describes the condition where bleeding occurs during the surgery. This is, strictly speaking, not an action as such, but should rather be considered an external condition. The actualisation of this condition may, however, be the consequence of the variability of a number of upstream functions, which are not described here.

**Table 9.10    FRAM representation of <Haemorrhage>**

| Name of function | Haemorrhage |
|---|---|
| Description | This is a condition that may occur during surgery |
| **Aspect** | **Description of aspect** |
| Input | *Not described initially* |
| Output | Uncontrolled bleeding |
| Precondition | *Not described initially* |
| Resource | *Not described initially* |
| Control | *Not described initially* |
| Time | *Not described initially* |

Since haemorrhaging is also considered to be a background function, only the Output is defined. None of the five other aspects are described any further.

In order to complete the first iteration of the FRAM model, only the description of <Complete surgical procedure> remains.

**Table 9.11     FRAM representation of <Complete surgical procedure>**

| Name of function | Complete surgical procedure |
|---|---|
| Description | To be done by the main surgeon and the assisting surgeon |
| **Aspect** | **Description of aspect** |
| Input | Tissue sample has been excised |
| Output | Surgical procedure has been completed |
| Precondition | Haemostasis has been achieved |
| Resource | *Not described initially* |
| Control | *Not described initially* |
| Time | *Not described initially* |

The three aspects that are described for this function have already been described by other functions. The first iteration of the FRAM analysis is therefore finished.

At this stage of the analysis, the consistency and completeness of the FRAM model should be checked. The consistency can be checked by making sure that aspects are described using the same names. If, for instance, one aspect is named [Surgical procedure has been completed] and another [All surgical procedure has been completed], then it should be ascertained whether this refers to the same aspect or to two different aspects. In the former case, the names should be adjusted to be identical. In the latter case, the difference between the aspects should be emphasised by making the names more distinct. The completeness can be checked by going through the functions one by one to make sure that all aspects that are described for one function can be found among the aspects of one or more of the other functions. In other words, there should not be any aspects that are 'loose' or left dangling in the air. This, of course, requires that the consistency has been checked first.

*The Sources of Variability*

Since essentially all the functions included in the FRAM model are carried out by humans, all the functions are potentially variable. The performance variability can be found by using descriptions of various sources of variability (see Chapter 6).

*Internal variability*
Since the duration of this event was rather short, it can be assumed that there was no internal variability of the human functions. In other words, the surgery did not last long enough for anyone to become fatigued by the surgery itself, nor were there effects of circadian rhythm. There may, of course, have been effects coming from work during the prior work periods, but in the absence of data about that it is assumed that internal variability was not significant. The same reasoning goes for the internal variability of the organisational functions, that is, the team in the main OR.

The haemorrhage that occurred during the surgery may be considered a special case of internal (physiological, anatomical) variability. Haemorrhage may happen for a number of reasons during abdominal surgery, including variability of the surgical procedure and the general state of health of the patient.

*External variability*
In this case an obvious source of variability was the double – or even triple – surgical procedures, that is, the surgery in the main OR plus the two other urgent surgical procedures carried out by the main surgeon and the assisting surgeon. This led to several phone calls to the main OR throughout the event. It also meant that both the main surgeon and the assisting surgeon were away, the first one time and the second two times. This contributed to a high workload situation, and also led to several interruptions of what should otherwise have been a smooth surgical procedure. In addition to that, the scrub nurse had recently qualified and was under supervision by the supervising nurse. The supervising nurse thus had more to do than usual. Finally, the patient took part in a study, which required a special analgesic to be administered towards the end of the surgical procedure. This also introduced variability in the procedure.

These conditions altogether make it reasonable to assume that there was variability in how the scrub nurse performed and that this was exacerbated by the additional work resulting from the discovery that the syringe had been prepared with the wrong analgesic. The main and the assisting surgeons may also both have been more variable in their performance since they were under considerable time pressure to complete their work so that they could attend to other surgical procedures.

Another consequence of this situation was that the communication to and from the main OR suffered, thereby creating another external source of variability. It was not the technical communication function as such that varied (that is, the quality of the communication), but rather that communication was not available because the main OR had only one phone, which was occupied, for instance, when the nurse anaesthetist needed it.

*Upstream–downstream coupling*
The variability of the general and specific inputs can be described by characterising the variability of the Output from the various functions, the assumption being that the variability of the Output is a consequence of the variability of the (performance of) the function. Using the simple solution to describe performance variability (Chapter 6), the Output can vary in terms of time and precision. Table 9.12 summarises the variability of the Output for the functions where performance for one reason or another varied. In the case of <Haemorrhage>, an elaborate term is used, since the presence of too much blood cannot otherwise be adequately described.

**Table 9.12    Output variability in the 'Duk i buk' case**

| Function | Output | Variability of output |
|---|---|---|
| Suturing the wound | Patient is ready for waking | Too late. Due both to the haemorrhage and the problems with the special analgesic |
| Count instruments and materials (BS) | All instruments and materials are accounted for | Too early. The scrub nurse made the count when some of them were still in the patient's abdomen |
| Prepare special analgesic | Syringe with special analgesic | Wrong object/wrong analgesic |
| Administer special analgesic | Special analgesic has been administered | Too late. The syringe had been prepared with the wrong analgesic, hence had to be replaced |
| Count instruments and materials (AS) | All instruments and materials are accounted for | Too late. It was carried out after the main surgeon had left the OR |
| Achieve haemostasis (stop bleeding) | Haemostasis has been achieved | Too late. The haemorrhage was more complicated than usual, and therefore took longer to control |
| Haemorrhage | Uncontrolled bleeding | Too much |
| Complete surgical procedure | Surgical procedure has been completed | Too late, due to haemorrhage |
| Waking the patient | *Not described initially* (but it could be [Patient is awake]) | Too early. Everyone was eager to complete the surgery due to the many disruptions and delays |

## An Instantiation of the Model

The potential couplings among the aspects are all contained in the FRAM model of the event, as is the potential variability of the functions. The actual variability can be described by considering an instantiation of the model, in this case corresponding to the actual event. The instantiation can, however, not be a complete description of what happened unless the model is developed further.

A number of aspects have not been described in the first iteration of the FRAM model. This is clearly something that must be done in a second iteration. Prime among those are the description of Time, since there was a considerable time pressure for reasons discussed above. This could be represented by having a function <Call surgeon from other OR> upstream to the T-aspect of <Complete surgical procedure>. Another important aspect is the Resource or Execution Condition, for instance the availability of the surgeons and the experience of the nurses. It may also be necessary in a second iteration to describe the Control aspect, for instance if the function for the preparation of the syringe is introduced when the <Prepare special analgesic> is elaborated upon. It is reasonable to assume that this was not a routine activity, hence prone to variability. I shall, however, leave a second iteration of the FRAM model to the interested reader.

For the purpose of illustration only, an instantiation of the first iteration of the FRAM model is shown in Figure 9.1 below.

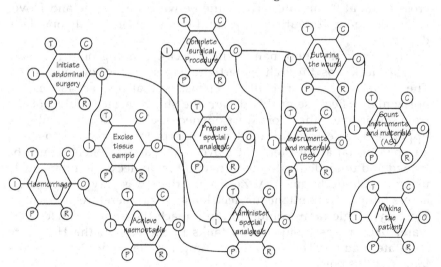

**Figure 9.1**     **Instantiation of the FRAM model for 'Duk i buk'**

The main problem in this event was that the counting/BS took place too early, and that no one seemed to pay any attention to it. The discrepancy was first discovered during the final counting/AS, which was after the wound had been sutured. The variability in the situation thus meant that the suturing function varied in the sense that the Precondition (counting) was not checked, and also that the waking up of the patient was started before the final count had been made.

### *Herald of Free Enterprise* Car Ferry Disaster

The second case is the capsizing of the roll-on roll-off (ro-ro) passenger and car ferry *Herald of Free Enterprise* outside the Belgian port of Zeebrugge on 6 March 1987. The vessel was en route to Dover in England but capsized soon after leaving the harbour, ending on her side half-submerged in shallow water, leading to the death of 193 passengers and crew.

#### *The Event as It Happened*

The *Herald of Free Enterprise* was a modern ro-ro passenger and vehicle ferry designed for use on the high-volume Dover–Calais ferry route and certificated to carry a maximum of 1,400 persons. The Dover–Calais route is very short, about 22 nautical miles, and the time to cross is about 90 minutes. The route between Zeebrugge and Dover is longer, about 72 nautical miles, and there is usually only one daily departure.

The *Herald* had two main vehicle decks. Loading onto the main vehicle deck was through watertight doors at the bow and stern. The doors were designed like a clamshell that opened and closed horizontally. Because of this design the ship's master could not see from the bridge if the doors were opened or closed. Loading onto the upper vehicle deck was through a weatherproof door at the bow and an open portal at the stern. At Dover and Calais, vehicles could be loaded and unloaded onto the upper and main decks simultaneously using double-deck ramps. At Zeebrugge there was only a single-level access ramp and simultaneous deck loading was therefore impossible. Since the single ramp could not quite reach the upper vehicle deck, water ballast was pumped into tanks in the bow of the Herald to facilitate loading. The turnaround time for the ferry was therefore longer at this port.

When the *Herald* left Zeebrugge, not all the water had been pumped out of the bow ballast tanks, causing her to be some three feet down at the bow. The assistant bosun was responsible for closing the bow doors. He had opened the bow doors on arrival at Zeebrugge and then supervised some maintenance and cleaning activities. After this he was released from this work by the bosun, and went to his cabin. He fell asleep and was not awakened by the 'harbour stations' public address call alerting the crew to take their assigned positions for departure from the dock.

The bosun left the car deck at the 'harbour stations' call to go to his assigned station. He later stated that it had never been part of his duties to close the bow doors or even make sure that anyone was there to do it. The chief officer was in charge of loading vehicles. He stated that he remained on the car deck until he thought he saw the assistant bosun threading his way through the parked cars toward the door control panel. The chief officer then went to the bridge, which was his assigned position for departure from dock.

The *Herald* left the berth stern first and then made a 180-degree starboard turn before leaving the inner harbour. She passed the outer mole and accelerated rapidly toward her service speed of 22 knots. As she increased speed, a bow wave began to build up under her prow. At 15 knots, with the bow down three feet lower than normal, water began to break over the main car deck through the open doors at a rate of 200 tons per minute. Owing to the lack of subdividing bulkheads in these types of vessels, water could easily flow from end to end and from side to side. The flood of water through the bow doors quickly caused the vessel to become unstable.

The *Herald* listed 30 degrees to port almost instantaneously. Large quantities of water continued to pour in and fill the port wing of the vehicle deck, causing a capsize to port 40 seconds later at 19.05 local time. The *Herald* settled on the sea bed at slightly more than 90 degrees with the starboard half of her hull above water. There had been no chance to launch any of the ship's lifeboats.

## The Functions and the FRAM Model

This event can be described in terms of three high-level functions that also represent the sequence of activities needed for the vessel during its departure from Zeebrugge:

- complete loading
- pre-sailing operations (close the bow port, de-ballast or restore vessel to natural trim)
- manoeuvring – sailing (sound harbour station call, man stations, drop moorings, leave harbour and accelerate to service speed

Even this general and non-technical description points to a number of clearly identifiable functions. They are, in sequence, <Complete loading>, <Close bow doors>, <Restore natural trim>, <Sound harbour station call>, <Man harbour stations>, <Drop moorings>, <Leave harbour> and <Accelerate to service speed>. The FRAM analysis starts by characterising these functions and how they are mutually dependent. This is in a sharp contrast to a conventional accident investigation, which would start with the capsizing and then go backward step by step until a set of necessary and sufficient causes had been found.

To illustrate that difference, the FRAM analysis can begin by looking at the function <Leave harbour>. With the benefit of hindsight, this function was carried out even though the Preconditions were not fulfilled. The analysis follows the same principles as the previous case, and the detailed comments will therefore not be repeated here.

Table 9.13    FRAM representation of <Leave harbour>

| Name of function | Leave harbour |
|---|---|
| Description | Vessel leaves the berth and sails through the inner harbour to start the sea voyage |
| **Aspect** | **Description of aspect** |
| Input | Harbour stations are manned |
| Output | Vessel has left harbour |
| Precondition | Vessel is trimmed |
| | Bow doors are closed |
| Precondition | Moorings have been dropped |
| Resource | *Not described initially* |
| Control | *Not described initially* |
| Time | Sailing schedule |

The description of the <Leave harbour> function points to seven other functions, one for the Input, one for the Output, three for the Preconditions, one for the Resource, and one for Time. Although the FRAM requires that all of these functions are described, this example will concentrate on a few of them.

The Output [Vessel has left harbour] is the Input of a function that can be called <Commence voyage>. This is not described further in this example. The Time input [Sailing schedule] can be seen as the

Output of a background function, which might be called <Produce winter schedule>. This is not described further in this example. The three Preconditions, however, need to be described further.

Table 9:14    FRAM representation of <Close bow doors>

| Name of function | Close bow doors |
| --- | --- |
| Description | Closing the bow doors before the vessel leaves harbour. This was assigned to the assistant bosun |
| **Aspect** | **Description of aspect** |
| Input | Assistant bosun at bow doors |
| Output | Bow doors are closed |
| Precondition | Chain in place |
|  | Harbour stations are manned |
| Resource | Loading Officer |
| Time | *Not described initially* |
| Control | Ship's Standing Orders |

The function <Close bow doors> points to six other functions. Three of these will only be mentioned, but not included in the short description given here. They should, however, be included even in the first iteration of constructing the FRAM model.

The Precondition [Chain in place] is the Output of a function that can be called <Finish loading>. (The chain is the physical chain that prevents vehicles from driving onto the car decks.) This function is not described further in this example.

The Resource [Loading officer], referring to the Loading Officer's supervision of the crew, is the Output of a function that could be called <Supervising crew>. This function is not described further in this example.

The Control [Ship's Standing Orders] refers to the instructions and procedures that regulate the normal activities on board the vessel. These instructions and procedures were presumably issued by the company that operated the ferry. This function is not described further in this example.

**Table 9.15    FRAM representation of <Trim vessel>**

| Name of function | Trim vessel |
|---|---|
| Description | This re-establishes the natural trim of the vessel. This can take considerable time |
| **Aspect** | **Description of aspect** |
| Input | Loading is completed |
| Output | Vessel is trimmed |
| Precondition | *Not described initially* |
| Resource | *Not described initially* |
| Time | *Not described initially* |
| Control | Ship' Standing Orders |

This is a relatively simple function, since it mainly involves aligning valves and starting the pumps. Once that has happened, the crew must wait for the pumping to be completed. In the description shown here, the Control refers to [Ship's Standing Orders] already described.

**Table 9.16    FRAM representation of <Drop moorings>**

| Name of function | Drop moorings |
|---|---|
| Description | Dropping the moorings is part of the departure procedure |
| **Aspect** | **Description of aspect** |
| Input | Harbour stations are manned |
| Output | Moorings have been dropped |
| Precondition | *Not described initially* |
| Resource | *Not described initially* |
| Control | *Not described initially* |
| Time | *Not described initially* |

The <Drop moorings> function is also described in a very simple manner. This completes the three Preconditions of <Leave harbour>.

**Table 9.17    FRAM representation of <Finish loading>**

| Name of function | Finish loading |
|---|---|
| Description | These are the operations that complete the loading, including putting the chain in place |
| **Aspect** | **Description of aspect** |
| Input | *Not described initially* |
| Output | Loading is completed |
| | Clear to call for harbour stations |
| Precondition | *Not described initially* |
| Resource | *Not described initially* |
| Control | *Not described initially* |
| Time | *Not described initially* |

Note that this function is described as having two different outputs. One is the signal that [Loading is completed]. The other is that the vessel is ready for the call for harbour stations.

**Table 9.18    FRAM representation of <Man harbour stations>**

| Name of function | Man harbour station |
|---|---|
| Description | This is the signal for the crew to take their positions for leaving the harbour |
| **Aspect** | **Description of aspect** |
| Input | Clear to call for harbour stations |
| Output | Harbour stations are manned |
| | Assistant bosun at bow doors |
| Precondition | *Not described initially* |
| Resource | *Not described initially* |
| Control | *Not described initially* |
| Time | *Not described initially* |

This function marks an important phase in the preparations for going to sea. The function has two different outputs. One is that harbour stations in general have been manned. The other is that the Assistant Bosun is at his position at the bow doors.

## The Sources of Variability

A main source of variability was that the vessel sailed to Dover from Zeebrugge rather than her normal route from Calais. At Zeebrugge the unloading and loading of cars took longer due to the lack of double-deck ramps. Since the company made no allowance for that, the more complicated procedure introduced a time pressure if the sailing schedule was to be maintained.

### Internal variability
If we consider the main technical functions, the bow doors were open, but not because they had malfunctioned. If the crew had tried to close them, the mechanism would (presumably) have worked. The pumps used to empty the ballast tanks also worked, but insufficient time was provided to empty the tanks so that natural trim could be re-established. Overall, there is nothing in the reports from the accident that suggests technological malfunctions.

In terms of the human functions, there was clearly some variability. Even though the accident happened in the late afternoon or early evening, the assistant bosun was sound asleep in his cabin. He was obviously tired, and this may have been due to his working hours, how long he had rested since the previous shift and so on. For the other main players involved (the bosun, the chief officer, the captain) it is plausible that the time pressure described above may have increased the variability of performance, for instance by tempting them to gain time by making trade-offs between efficiency and thoroughness.

For the organisational functions, it can be assumed that there was no variability during the event. Indeed, the whole event only took about one hour, with the actual duration of sailing being about 20 minutes. The frequency of the variability of organisational functions is usually orders of magnitude lower than that.

### External variability
For the technological functions, there is no indication that these did not perform as required. (In this event, one or more technological functions were missing, such as an indication of the status of the bow doors. But that is a different matter.)

In the case of the *Herald*, there were several external sources of variability for the human and organisational functions. For the human functions there was the pressure to keep the schedule, that is, to arrive and leave on time. Several functions were truncated in order to gain time, either by starting them prematurely or by taking things for granted. This had consequences both for the de-ballasting of the vessel, which had not been completed when she left the harbour, and for how people behaved. In several cases people neglected to check something, presumably because they believed that it was not necessary. The chief officer left the main deck to go to his harbour station on the bridge because he thought he saw the assistant bosun moving towards the bow doors. The captain worked on the assumption that the bow doors were closed, unless he had been told otherwise. The time pressure also influenced the decision to sail before the natural trim had been re-established, on the assumption that it was safe enough to do so.

Another influence was the 'spirit' or culture in the organisation – both on board the vessel and in the operating company. This is illustrated by the bosun's statement that it was not his duty to close the bow doors. This can be seen as an emphasis on being efficient in one's own work without being concerned about what happens around.

*Upstream–downstream coupling*
From a nominal point of view, little of the variability in this event was due to the couplings among general and specific inputs. The reason why the vessel was not de-ballasted before it began to sail was not that the de-ballasting took longer than usual, but rather that not enough time was allowed for it to be completed. The internal variability of several of the human functions was generally due to the time pressure, which made it important to complete stages of work in time, hence to take short-cuts.

**Table 9.19  Variability of Outputs for the '*Herald*' case**

| Function | Output | Variability of output |
|---|---|---|
| Leave harbour | Vessel has left harbour | Too early, in the sense that the vessel was not ready to go to sea. Too late, in the sense that it was behind schedule |
| Close bow doors | Bow doors are closed | Omission. The bow doors were not closed |
| Man harbour stations | Harbour stations are manned | Imprecise or incomplete. Not all harbour stations were manned as they should have been |

The natural trim of the *Herald* had not been established when she started her voyage. This was, however, not because the function <Trim vessel> varied, for instance in the sense that the output was delayed.

Emptying the ballast tanks was a process that took its time. Although it had not been completed when the Herald left her berth, that was because <Leave harbour> was started too early, rather than because <Trim vessel> ended too late.

For two of the other functions included in the first iteration of the model, neither <Finish loading> nor <Drop moorings> is assumed to have shown any significant variability.

### An Instantiation of the Model

An instantiation of the FRAM model illustrates how the variability described in Table 9.19 can be used to understand how the accident happened. The main variability in this case was a strongly perceived need to start sailing, which meant that some of the Preconditions were not verified as they should have been. The captain habitually relied on negative reporting, which means that he assumed that the bow doors were closed unless he was explicitly told that they were open. The officers on the bridge may also have assumed that it was safe to leave the berth even though the vessel was not in trim; after all, the bow doors were assumed to be closed. Finally, the <Man harbour stations> also varied, in the sense that the Output was imprecise, and perhaps even premature.

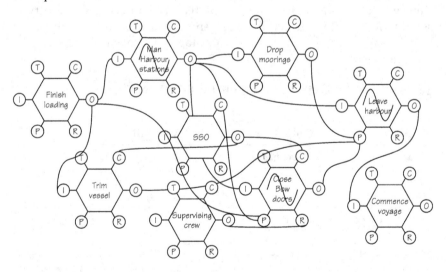

**Figure 9.2        Instantiation of the FRAM model for the 'Herald' case**

This first iteration of building the FRAM model of the 'Herald' case is clearly rather rough and incomplete. But it illustrates how a model can be constructed, and also how it can be elaborated. Each of the functions can – and should – be described in further detail, which will lead to a model that more precisely describes what should have happened. This can then be used as a basis for an improved understanding of what did happen.

## A Financial System

The last case shows how the FRAM can be used to analyse a financial system, or rather some basic functions of a financial system. Most people are still painfully aware that financial systems can get out of control, with unanticipated, significant, and widespread consequences for almost everyone. The purpose of this last case is not to present a FRAM analysis of the financial crisis in the USA in 2007–2008, but rather to illustrate how the FRAM can be used to identify and represent the basic functions of a financial system so that it becomes easier to understand the role of increased variability and to identify the possible consequences.

### The System under Consideration

The case is not about any specific or named system but rather about some of the obvious functions in the general financial market. Perhaps the most important purpose of a financial system, such as a bank, is to consider requests for funding, for instance to buy a home or finance an industrial activity, and to grant the requested loan if the necessary conditions are met. This is a situation that most people have experienced at one time or another, whether the request has been for a small loan to buy a new car or for large loan to buy a house.

A typical sequence of events would be that a requestor, say an individual, applies for a loan to a financial institution (a bank). In order for the bank to decide whether to grant the loan, it is necessary to evaluate the credit-worthiness of the requestor, which essentially is a financial risk analysis. If the outcome is positive, the bank can then transfer the money to the requestor (after the proper papers have been signed, of course, and provided the bank has the funds). While the loan is in force, the bank must continue to manage the risk of the loan, since the value of the collateral, such as a house, may be reduced if market conditions deteriorate. In the same way, a regulator must manage the risks of the market, for example, the banks, to make sure that proper procedures are followed. The banks may rely on their own capital or may from time to time need additional capital, which in turn may depend on the market conditions and so on. As even this brief and non-technical description makes clear, there are a number of functions to consider in developing a model of a financial system.

## The Functions and the FRAM Model

Since this analysis looks at a generic financial system rather than a specific event there is no 'natural' starting point; one must therefore be chosen. In this example it will be a function called <Transfer funding to requestor>. This is an essential function of the financial system, and in some sense the very reason for its existence. Starting from this function, the FRAM can be applied, as described in Chapter 5. As outlined above, there must clearly be an Input to this function, that is, something that starts the transfer. There must also be at least one Precondition, since loans should not simply be approved for anyone who requests them. There must also be a Resource which provides the funding that is transferred. And there may possibly be both Time and Control inputs to regulate the transfer, although these need not be described from the start.

Table 9.20    **FRAM representation of <Transfer funding to requestor>**

| Name of function | Transfer funding to requestor |
|---|---|
| Description | This is a function carried out by the financial institution (bank) |
| Aspect | Description of aspect |
| Input | Approved amount |
| Output | Loan |
| Precondition | Requestor has been approved |
| Resource | *Not described initially* |
| Control | *Not described initially* |
| Time | *Not described initially* |

This description of <Transfer funding to requestor> points to three other functions. One is needed to provide the specific Input [Approved amount]. A second is needed to provide the Precondition [Requestor has been approved]. And a third is needed as the receiver of the Output [Loan]. While the two upstream functions must be described further, we shall conveniently assume that the third is a background function or sink, for instance as a <Purchase> function, that does not need to be elaborated.

Bringing about the [Approved amount] can be represented by a function called <Provide funding>, described as follows:

**Table 9.21    FRAM representation of <Provide funding>**

| Name of function | Provide funding |
|---|---|
| Description | This is a function carried out by the financial institution (bank) |
| **Aspect** | **Description of aspect** |
| Input | Approved request for loan |
| Output | Approved amount |
| Precondition | *Not described initially* |
| Resource | Economic resources |
| Control | Risk management guidelines |
| Time | *Not described initially* |

The description of <Provide funding> in turn points to three other functions. One is needed to provide the Input [Approved request for loan]. A second is needed to provide the Resource [Economic resources]. And a third is needed to provide the Control [Risk management guidelines]. The output, [Approved amount] has already been described by another function.

For the sake of simplicity, we will assume that the Resource [Economic resources] is the Output of a background function that does not need to be described further for the example. (But it is clearly a function that can vary, as the recent financial crisis has demonstrated.)

The <Transfer funding to requestor> is contingent upon a satisfactory evaluation of the requestor's creditworthiness. This can be represented by a function called <Credit risk assessment>, described as follows:

**Table 9.22    FRAM representation of <Credit risk assessment>**

| Name of function | Credit risk assessment |
|---|---|
| Description | This is a function carried out by the financial institution (bank), or possibly a specialised service provider |
| **Aspect** | **Description of aspect** |
| Input | Approved request for loan |
| Output | Requestor has been approved |
| Precondition | *Not described initially* |
| Resource | *Not described initially* |
| Control | Risk management guidelines |
| Time | *Not described initially* |

Both <Provide funding> and <Credit risk assessment> point to two other functions. One is needed to provide the Input [Approved request for loan], while a second is needed to provide the Control [Risk management guidelines]. These can be described as follows:

**Table 9.23    FRAM representation of <Process request>**

| Name of function | Process request |
|---|---|
| Description | This is a function carried out by the financial institution (bank). It is the administrative procedure in response to a request for a loan |
| **Aspect** | **Description of aspect** |
| Input | Request for loan |
| Output | Approved request for loan |
| Precondition | *Not described initially* |
| Resource | *Not described initially* |
| Control | *Not described initially* |
| Time | *Not described initially* |

For the sake of keeping the example simple, this function 'stands on its own', so to speak, although it clearly is possible to develop it further by considering, for example, the Time and Control aspects. The same goes for the <Risk management> function. Both are therefore initially treated as background functions.

**Table 9.24    FRAM representation of <Risk management>**

| Name of function | Risk management |
|---|---|
| Description | This is a function carried out by official (national or international) regulators of the financial markets |
| **Aspect** | **Description of aspect** |
| Input | *Not described initially* |
| Output | Risk management guidelines |
| Precondition | *Not described initially* |
| Resource | *Not described initially* |
| Control | *Not described initially* |
| Time | *Not described initially* |

Even this simplified FRAM model illustrates several possible couplings among the functions. Thus a demand can initiate the provision of economic resources as well as a risk assessment of the requestor. The availability of the necessary resources can be a 'signal' to initiate the transfer, provided there is a satisfactory result of the credit risk assessment. Finally, both the provision of funding and the credit risk assessment are guided or controlled by independent risk management.

The couplings describe how functions may affect each other, not in terms of causes and effects but in terms of propagation of variability. If, for instance, <Credit risk assessment> does not work satisfactorily (for example, that either anticipation or monitoring do not work well), then <Transfer funding> may be adversely affected.

## The Sources of Variability

As recent events have shown, and to some extent continue to show, there is abundant variability in how the financial markets perform. One may indeed suggest that the financial markets are unlike other processes in the sense that they do not fail as such, but rather function with different degrees of efficacy (and acceptability). Although the financial market can be dysfunctional, it cannot capsize like the *Herald of Free Enterprise* in the sense that it reaches an irreversible state. (The same, however, does not go for individual companies.)

### Internal variability
For technological functions we shall remain with the default assumption that the variability of technological functions is low, the 'Flash Crash' in May of 2010 notwithstanding. For human functions, the internal variability is assumed to be high, mainly because of psychological factors. (This is so even when the example is something as mundane as processing of loans.) One source of performance variability is definitely the tendency to rely on cognitive trade-offs. The variability of organisational functions is probably low under 'normal' conditions and should indeed be so to allow the organisation to act as a regulator or attenuator of individual variability. But variability may be high when the internal functioning of the financial institution occasionally is tumultuous.

### External variability
As far as the external variability is concerned, this exclusively affects human and organisational functions. Processing individual requests for loans, as well as managing the overall loan mass for a financial institution, can be a delicate case of management that is influenced

by numerous external factors. These include the market dynamics – partly unpredictable and partly unknown, a highly competitive environment, uncertainties about how the markets may develop both in the short and the long-term, and a dependency on 'irrational' psychological factors as well as political developments that are outside the control of the financial system.

*Upstream–downstream coupling*
In the financial system, this source of variability is assumed to be the most important. The financial system must function in a rather turbulent environment, and what it does will to a large extent determine what happens in that environment. This means that a model of the financial system should be considerably larger than the example used here. There are a number of dependencies or couplings that may affect both the Resource, Control and Time aspects of functions and thereby easily lead to unstable dynamics. One could actually argue, with considerable justification, that the financial system should not be seen as separate from the society with which it interacts and vice versa.

**Table 9.25     Variability of Outputs for the financial case**

| Function | Output | Variability of output |
|---|---|---|
| Transfer funding to requestor | Loan | Too early. The transfer may happen before the credit risk assessment has been completed, in anticipation that it will be a positive one |
| Credit risk assessment | Requestor has been approved | Imprecise. The credit risk assessment may be unreliable |
| Process request | Approved request for loan | Imprecise. Requests may be approved without sufficient scrutiny |
| Risk management | Risk management guidelines | Imprecise. The guidelines may be too vague, or not cover all aspects of risk management |

The function <Provide funding> is the only one that is not initially assumed to vary. The reason is that this is basically an administrative function, where the approved request triggers the release of the [Approved amount] from the economic resources and transfers it to <Transfer function>.

## An Instantiation of the Model

Since this case describes a possible future situation rather than an event that has happened, it is limited by the assumptions about the variability of the functions. One approach would be to take an analytic stance, and consider how the potential variability of each function could become actual variability. This would be manageable for the first iteration of the model described above, but may be less attractive for more elaborate versions that easily may contain 30–40 functions or more.

Another approach would be to look at the lessons learned from major financial crises, including but not being limited to the most recent one. Here experience shows that the performance of functions such as <Risk management>, <Credit risk assessment> and <Process request> can vary quite a lot, to the extent that they almost become token functions.

For the purpose of illustration only, an instantiation of the first iteration of the FRAM model is shown in Figure 9.3 below.

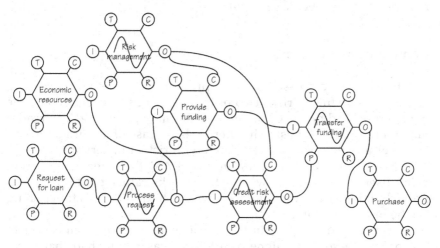

**Figure 9.3    An instantiation of the FRAM model for the Financial case**

The efficacious performance of the financial system requires that loans are not provided unless it is prudent to do so. This means both that the credit risk assessment works as intended, and that the transfer of funding works as intended. But if we imagine that the credit risk assessment becomes too variable (for instance, too superficial), then loans will be granted when they should not have been. This can in principle be caught by the Risk management (the regulators), but if their performance also is variable, in the sense that it is not sufficiently thorough or strict, then the flow of money will begin to increase. This may potentially lead to a situation that is out of control. (There is here a coupling between the ease by which loans are granted and the size of requests for loans. This is, however, not a simple proportional relation, but involves the relation to other factors such as the stability of society, the economic outlook, the stability of the past years and so on.)

An increase in the number of loans will require easier access to funding. A second iteration of the model should therefore expand on the <Supply of money> function and how this may vary. The recent financial crisis can easily provide some suggestions. A more detailed description of the timing of the functions will also be important, since the financial markets are very volatile. For reasons of space it is impossible to do so here. The illustration has hopefully achieved its purpose by suggesting how the FRAM can provide an understanding of how the financial markets function, and how performance variability may play out in specific scenarios.


## Comments on the Cases

The three cases discussed above represent an incident, an accident, and a possible future risk. Two of them are thus of events that have gone wrong and one is of something that may go wrong. The reader may therefore rightly ask why there was no case of a 'normal' event, of something that was not associated with risk or uncertainty.

There is, of course, nothing in the FRAM that precludes it from being used to describe how a system works rather than how it can possibly fail – or even to find ways in which the system's functioning can be improved, for example, with regard to productivity or quality. It would indeed be quite in the spirit of Resilience Engineering to do so. The reason why none of the three cases have had that focus is simply convenience, since it is much easier to find descriptions of things that have gone wrong than of things that have gone right.

Having said that, it should nevertheless be pointed out that the FRAM model in all three cases is a model of everyday performance, hence a description of how each system functions – or should function – to serve its purpose. The instantiations are of situations or conditions when something failed, or might fail. But it is equally possible to create instantiations of situations or conditions where everything functions as it should. This could be used as a basis for thinking about productivity and quality rather than just safety – in other words about resilience in its most general form.

## Comments on Chapter 9

The descriptions of the three cases have made use of existing descriptions as well as FRAM analyses in whole or in part.

The medical case, 'duk i buk,' has been described in a MSc thesis by Helen Alm, and also in Alm, H. and Woltjer, R. (2011), 'Patient safety investigation through the lens of FRAM', in D. de Waard, A. Axelsson, M. Berglund, B. Peters and C. Weikert (eds), *Human Factors: A System View of Human, Technology and Organisation*, Maastricht, the Netherlands: Shaker Publishing. I am grateful for the permission of Helen Alm to use her work.

The *Herald of Free Enterprise* has been the subject of several *post mortem* examinations, including the formal investigation by Lord Justice Sheen. For the FRAM analysis I have made use of material in Praetorius, G., Lundh, M. and Lützhöft, M., 'Learning from the past for pro-activity – A re-analysis of the accident of the MV Herald of Free Enterprise', which was presented at the Fourth Resilience Engineering Symposium in June 2011, and also had the benefit of the advice from both Gesa Praetorius and Margareta Lützhöft.

The financial case is based on a more extensive treatment in Sundström, G.A. and Hollnagel, E. (eds.), (2011), *Governance and Control in Financial Systems: A Resilience Engineering Approach*, Farnham, UK: Ashgate. Gunilla Sundström has done her best to make me understand how the world of finance works.

The 'Flash Crash' mentioned in this chapter refers to the few minutes on 6 May 2010 during which the Dow Jones Industrial Average plunged about 900 points and then recovered.

# Chapter 10
# Afterthoughts

We are too much accustomed to attribute to a single cause that which is
the product of several, and the majority of our controversies come from
that.

(Marcus Aurelius)

## The FRAM: A Method But Not a Model

A method always embodies a set of underlying principles that
encapsulate the salient features of the phenomena that it can address.
While the principles on which a method is based need not be described
directly as a model, it is nevertheless often the case, not least in safety
studies – accident investigations and risk assessments. The models
are simplified representations either of how accidents can happen,
or of how the system being described is organised – usually in terms
of some kind of hierarchical ordering of layers, parts or components.
Because the model defines what the method can be used for, it also
defines the limits of the method.

The underlying principles are sometimes clearly described but are
more often taken for granted, either because they refer to 'common
sense', or because the method has been in use for so long that the
principles no longer are considered, let alone challenged. Even when
the principles behind a method are described, the descriptions are not
always clear. Consider, for instance, this description of Root Cause
Analysis (RCA):

> RCA assumes that systems and events are interrelated. An action in one area
> triggers an action in another, and another, and so on. By tracing back these
> actions, you can discover where the problem started and how it grew into the
> symptom you are now facing.

It is interesting to compare this to the First Axiom of Industrial Safety, which states that 'The occurrence of an injury invariably results from a completed sequence of factors – the last one of these being the accident itself.' The similarity is striking. RCA thus corresponds closely to the idea of the Domino model, according to which events develop as a chain reaction of causes and effects where the first domino piece that falls represents the original or root cause, while the last domino piece represents the final outcome or injury.

While the Domino model today is used rather loosely, in the sense that a domino piece just represents something that can fail, the original version of the Domino model defined five different pieces, called 'ancestry', 'person', hazard', 'accident' and 'injury' respectively. It also prescribed how an accident investigation should be carried out, starting by the 'injury' and going back step by step until the level of 'ancestry' was reached. The RCA is less constrained since it does not make an assumption about five distinct levels (that is, five different dominoes). On the other hand this makes it more arbitrary when the 'root' cause has been found – that is, what the stop rule of the analysis is.

It can, of course, be a significant advantage if methods refer to an explicit model. Such methods simplify safety analyses by offering a general description of how the world is structured or organised. If the description is correct, it clearly saves a lot of time. But if it is not, the method may produce results that are ambiguous, irrelevant or directly wrong.

## Methods That Rely on Strong Models

The above point can be illustrated by taking a closer look at three well-known methods, chosen because they are among the best known rather than because they are 'easy targets'. Each method provides a distinct model of how accidents happen, and each model-cum-method has strengths as well as weaknesses.

### AcciMap

The purpose of the AcciMap method is to identify the factors that caused (or failed to prevent) an accident. The AcciMap maps the possible causes of accidents onto six levels, ranging from 'Government, Policy and Budgeting' as the highest, to 'Equipment and Surroundings' as the lowest. The AcciMap analysis method tries to identify causal factors from all levels of the system in which the accident took place, usually by working its way from the physical sequence of events and activities right up to the causes at the governmental, regulatory and societal levels. The method refers to the notion of 'the anatomy of an accident', which is a kind of simplified fault tree that goes through five steps from root cause to target or victims. The similarity to the Domino model is obvious, despite the intervening 70 years. The main difference between the Domino model and AcciMap is that the latter refers to the larger socio-technical system described as a hierarchy of levels.

The purpose of the AcciMap analysis is to develop an understanding of the broad situational context within which the sequence of events took place, by repeatedly asking 'why' for each causal factor uncovered. In doing this, the method is similar to many others, such as the 'why-because' method, or even RCA. The prescription of the six levels obviously facilitates both the analysis and the graphical rendering of the results. But the six levels also greatly constrain the analysis by limiting it to these levels, by narrowing the terminology, and by making it difficult (although strictly speaking not impossible) to describe causal influences that do not correspond to the ordering of the levels. Finally, while the underlying model, the six levels, may be a plausible way to describe some industrial systems, it cannot be assumed to be generally applicable.

## Tripod

The Tripod approach to accident analysis targets the underlying problems that lead to incidents. The Tripod theory emphasises that the immediate causes (unsafe acts, people's errors), do not occur in isolation but are influenced by external factors – organisational or environmental preconditions. Many of these factors originate themselves from decisions or actions taken by planners, designers or managers that are far away from the scene of the accident. The consequences of these actions are described in Tripod as latent failures. Since latent failures according to their nature can be expected to have an impact on a broad front, identifying and addressing them will bring wider benefits than simply identifying the immediate cause(s) of the accident.

A Tripod accident analysis focuses initially on the accident mechanism (the facts concerning the events and their consequences, including the potential consequences) and uses it as a structure to identify the hazard management measures (the controls and defences) that should have been in place. According to the Tripod theory, incidents happen because hazard management measures either failed or were missing. The investigation and analysis therefore examine the immediate and latent failures behind each missing or failed hazard management measure, and propose action items to address them. Tripod clearly embodies a sequential model that describes how several parallel sequences eventually can merge to produce the final outcome. It is also a causal model, in the sense that the lines between the boxes represent some form of causality.

*STAMP*

STAMP (Systems-Theoretic Accident Model and Processes) is an accident analysis method that explains accidents as a result of inadequate control or inadequate enforcement of safety-related constraints rather than as a result of component failures and malfunctions. It is thus clearly different from both AcciMap and Tripod. STAMP views safety as a control problem: accidents happen if internal or external events and disturbances are not adequately handled or controlled. This may move the system beyond the safe states implied or defined by the safety constraints, which specify the relationships among system variables or components that constitute the non-hazardous or safe system states. Safety is correspondingly seen as an emergent property that arises from system components interacting within an environment. STAMP emphasises the dynamics of how safety is controlled, and does not see the system as a static structure.

STAMP is nevertheless basically a linear causal analysis method based on a systems theory model that encapsulates a number of assumptions about how systems generally are structured. The model represents how components are organised (in a hierarchy) as a basis for understanding or depicting unanticipated interactions among components. The properties that emerge from a set of components at one level of hierarchy are controlled by constraining the degrees of freedom of those components, hence limiting system behaviour to the safe changes and adaptations implied by the constraints.

## Articulated Model or Articulated Method?

As this discussion has shown, all safety analysis methods refer to a model, and most methods refer to an articulated model. The advantage of this is that the method basically prescribes a way of 'mapping' or understanding events – past or future – vis-à-vis the model. Past events are mapped onto the model, in the sense that they are explained by applying the assumptions of the model (levels, hierarchies, relations, components). Similarly, future events are developed by populating the model with specific details and deriving the consequences from that.

The difference between a model and a method can clearly be seen by comparing the FRAM to each of the three approaches described above.

- AcciMap refers to a distinct model that basically is an abstraction hierarchy of causes, with six levels ranging from 'equipment and surroundings' to 'government, policy and budgeting'. The method analyses accidents – and other events – using this model, and essentially maps the events onto the model.
- The purpose of Tripod is to describe how accidents happen using active and latent failures, and dysfunctional barriers. Tripod thus imposes a specific description of accidents. Tripod also relies on causal propagation, where the origin is found in failures and malfunctions.
- The generic version of STAMP is a model of socio-technical control comprising two hierarchical control structures, one for system development and one for system operations, as well as the interactions between them. The two hierarchies are essentially abstraction hierarchies, similar to the one used by AcciMap, going from the operating process at the lowest level to political decisions at the highest. The method serves to describe the actual events in terms of these control structures.

In all three cases, as in most other approaches to accident analysis, the underlying model defines or describes a set of relations while the associated method provides a way to interpret events in terms of those relations. The relations typically invoke the principle of causality (causes leading to effects or effects being preceded by causes), where the causes typically are failures or malfunctions of components, of functions, of control and so on. Since the models provide a clearly structured representation of the 'world', the methods are typically linear with either single or multiple cause–effect paths.

In these approaches, the methods in practice impose an a priori interpretative structure on the event. The value of the results of an analysis therefore depends on the correctness of the model, in particular whether it has been verified or validated. One might argue, and possibly even hope, that the correctness issue would be resolved in practice by virtue of the fact that a model with incorrect assumptions would disappear over time. However, people are adept at explaining away any such differences as demonstrated by an abundance of examples, usually by introducing convenient, supplementary ad hoc assumptions.

In everyday practice, which means in the short-term, the advantage of an articulated model is the efficiency of the associated method – even if the model is incorrect. The increased efficiency often outweighs the disadvantages, in particular the lack of thoroughness that is a consequence of the simplified model assumptions. Here, as in many other places, the ETTO principle wins hands down. However, in the longer term the opposite trade-off is more important.

Where commonly used methods try to describe relations derived from the model, and hence represent a model-cum-method approach, the FRAM can best be described as doing the opposite. The FRAM proposes that everyday events and activities can be described in terms of the functions involved without predefining any specific functions or assuming they are organised in a specific way. The FRAM further proposes that functions can be characterised using the six aspects, but does not rule out that a different set of aspects being used. The FRAM does not refer to a model, not even a non-linear one, and makes no assumptions about how the system under investigation is structured or organised, or about possible causes and cause–effect relations. The FRAM thus uses a method to produce a model, instead of a model to produce a method, and can therefore be described as a method-sine-model approach.

The FRAM is based on the four principles described in Chapter 3, but does not include any hidden assumptions. This means that if there is something in common among the models that the FRAM produces, it is simply because the cases or situations are similar. Within any given technological civilisation, processes and industries will be very similar in the way they are built and operated and the FRAM models will inevitably reflect that. But the similarity among the models is not a consequence of model assumptions as in the method-cum-model approaches, such as a hierarchy of levels, causal dependencies and so on. Indeed, the main purpose of the FRAM is to build a model of how a set of activities is carried out (and to create instantiations from that model), rather than a model of the system in the traditional sense. This also means that the method must be more detailed than most other 'methods', since the procedure of developing the model must stand on its own.

While the advantage of the FRAM is that it is thorough, as described above, the disadvantage is that it may require more work or effort to use the FRAM and that the efficiency therefore seems to be lower than for conventional methods. But as the reflections on the ETTO principle demonstrate, the trade-off between efficiency and thoroughness depends heavily on what the time horizon is. That which is more efficient in the short run – because it is faster to do – may in actual fact be less efficient in the long run, because it is less thorough. Less thoroughness means being less prepared or less ready, so that more effort will be needed later on, hence reducing efficiency.

The FRAM does not imply that events happen in a specific way, or that any predefined components, entities or relations must be part of the description. Instead it focuses on describing what happens in terms of the functions involved. These are derived from what is necessary to achieve an aim or perform an activity, hence from a description of work-as-done rather than work-as-imagined. But the functions are not defined a priori nor necessarily ordered in a predefined way such as a hierarchy. Instead they are described individually, and the relations between them are defined by empirically established functional dependencies (upstream–downstream couplings) rather than by the assumptions of an underlying model.

## Scale Invariance

One feature of the FRAM is that the description of functions is scale invariant. This basically means that the processes that give rise to phenomena on a small or a large scale are of the same kind. (The best known example of scale invariance is provided by fractal geometry.) If, for instance, we look at things from the point of view of failure, then there would be no difference between high-level failures or low-level failures – not least because there are no levels in a FRAM model. Specifically, there would be no need to have theories of organisational failure that were different from theories of human failure.

Few of the commonly used approaches are scale invariant, as the three examples above illustrate. This is because they refer to an abstraction hierarchy. It is an unspoken consequence of an abstraction hierarchy that processes – and therefore also failures – are different at different levels. The reason is that a model, by definition, cannot be scale invariant, since the very structure of the model imposes a scale: there are a top and a bottom and a number of layers in between.

A method, on the other hand, can be scale invariant if it is based on general principles and not linked to a specific model. The obvious advantage of scale invariance is the simplicity of the method and the parsimony of explanations. The latter is perhaps that most important, because it eliminates the need of large sets (or even taxonomies) of categories, that both can be cumbersome to use and that artificially constrain the depth and breadth of an analysis.

## Comments on Chapter 10

As the name implies, Root Cause Analysis (RCA) is an accident investigation method that aims to find the root cause, or root causes, of a problem on the assumption that if the root cause is fixed, then the problem will not recur. There many versions of the method, but they all follow the same principle. RCA is attractively simple and widely used, but ill-suited to anything but simple, homogeneous activities.

The 10 Axioms of Industrial Safety are described in Heinrich, H.W. (1959), *Industrial Safety* (4th edn), New York: McGraw-Hill. This edition of the book, which originally was published in 1931, also describes the famous Domino model of accident causation.

The AcciMap was literally intended to be a map of an accident, but the term is now generally used to include also the associated method. The original description can be found in Rasmussen, J. and Svedung, I. (2000), *Proactive Risk Management in a Dynamic Society*, Karlstad, Sweden: Swedish Rescue Services Agency. One of the precursors to the method is the so-called 'anatomy of an accident,' which has been described in Green, A.E. (1988), 'Human factors in industrial risk assessment – some early work', in L.P. Goodstein, H.B. Andersen and S.E. Olsen (eds), *Task, Errors and Mental Models*, London: Taylor & Francis.

The Tripod approach is based on the work of James Reason. It exists both in a reactive and a proactive version, named Tripod Beta and Tripod Delta respectively. Tripod was first described in Reason, J., Shotton, R., Wagenaar, W.A., Hudson, P.T.W. and Groeneweg, J. (1989), *Tripod: A Principled Basis for Safer Operations*, The Hague: Shell Internationale Petroleum Maatschappij.

The Systems-Theoretic Accident Model and Processes (STAMP) approach has been developed by Nancy Leveson. It was first presented in 2002 as a paper entitled 'A Systems Model of Accidents' at the International Conference of the System Safety Society in Denver, Colorado.

The FRAM itself was first called the Functional Resonance Accident Model, and is described as such in the *'Barriers'* book from 2004. It, however, soon became clear that the FRAM was a method rather than a model, and the meaning of the acronym was therefore changed. Since the FRAM proposes a specific data structure, it may still be considered a model from a software engineering perspective. But that is not the kind of model that safety studies are concerned about.

# Chapter 11
# FRAM on FRAM

'I don't see much sense in that,' said Rabbit.

'No,' said Pooh humbly, 'there isn't. But there was going to be when I began it. It's just that something happened to it along the way.'

(A.A. Milne, Winnie the Pooh)

The method itself has been described in this book in terms of one plus four major steps. By way of summary, and as a practical guidance, each step of the method is in the following described more succinctly. The obvious way to do that is to use the FRAM to describe itself – in other words, by describing each step as a function rather than as an action (or set of actions). To follow the main principles of the FRAM, the description will be textual rather than graphical. The reader may, as a complementary exercise, attempt to produce a graphical rendering of the FRAM model of the FRAM.

The five tables that follow define the five man functions of the FRAM. Each function contains references to other functions, but for reasons of space the other functions have not been developed here. But the description of the five main functions show that it is possible to make a FRAM analysis of the FRAM. Doing so would provide a good example of a FRAM model of how an activity should succeed, rather than of how it could fail.

**Table 11.1    A FRAM description of Step 0**

| Name of function | Step 0 Determine purpose and scope of the analysis |
|---|---|
| Description | The initial step sets the scene for the analysis. Although the following steps are similar, there may be differences with regard to needs, resources and so on |
| **Aspect** | **Description of aspect** |
| Input | Request for accident/incident investigation |
| | Request for risk assessment/risk analysis |
| | Need to understand everyday performance |
| | Need to assess possible effects of a change (design, innovation, remedial action) |
| Output | Work order |
| | Estimate of required resources (time, staff, data, materials and so on) |
| | Tentative analysis/investigation schedule |
| Precondition | Clear mandate from stakeholder(s) |
| Resource | *Not defined initially* |
| Control | *Not defined initially* |
| Time | *Not defined initially* |

**Table 11.2    A FRAM description of Step 1**

| Name of function | Step 1 Identify and describe the functions |
|---|---|
| Description | This step identifies the functions that are required for everyday work in the situation(s) being investigated |
| **Aspect** | **Description of aspect** |
| Input | Work order |
| | Detailed data about the target activity |
| Output | A FRAM model of the target activity |
| | Estimate of required resources for Step 2 |
| | Investigation/analysis schedule update-1 |
| Precondition | Full access to data sources and involved field personnel |
| Resource | FRAM expertise |
| | Subject matter (domain) expertise |
| Control | The FRAM (the method itself) |
| | Consistency/completeness check of model |
| Time | Tentative analysis/investigation schedule |

**Table 11.3    A FRAM description of Step 2**

| Name of function | Step 2 Describe potential and actual performance variability |
|---|---|
| Description | This step assesses the potential and actual variability of the functions in the FRAM model and in the instantiation(s), respectively. This should consider everyday variability as well as possible cases of excessive ('out of range') variability |
| **Aspect** | **Description of aspect** |
| Input | A FRAM model of the target activity |
| Output | Instantiation(s) of the FRAM model |
| | Investigation/analysis schedule update-2 |
| Precondition | *Not defined initially* |
| Resource | FRAM expertise |
| | Subject matter (domain) expertise |
| | General safety (human factors) experience |
| | Specific data/data bases if available |
| Control | The FRAM (the method itself) |
| Time | Investigation/analysis schedule update-1 |

**Table 11.4    A FRAM description of Step 3**

| Name of function | Step 3 Aggregation of performance variability |
|---|---|
| Description | This step describes the possible functional upstream–downstream couplings of variability, as a basis for understanding how variability can either increase (resonance) or decrease (damping) |
| **Aspect** | **Description of aspect** |
| Input | Instantiation(s) of the FRAM model |
| Output | Assessment of possible outcomes for a given instantiation |
| | Investigation/analysis schedule update-3 |
| Precondition | *Not defined initially* |
| Resource | FRAM expertise |
| | Subject matter (domain) expertise |
| | General safety (human factors) experience |
| Control | The FRAM (the method itself) |
| Time | Investigation/analysis schedule update-2 |

**Table 11.5    A FRAM description of Step 4**

| Name of function | Step 4 Propose ways to control variability |
|---|---|
| Description | This step develops recommendations for remedial actions, including monitoring (KPI), barriers and facilitators |
| **Aspect** | **Description of aspect** |
| Input | Assessment of possible outcomes for a given instantiation |
| Output | Recommendations for remedial actions |
| | Specification of performance indicators (KPI) |
| | Recommendations for effective control strategies |
| Precondition | *Not defined initially* |
| Resource | Subject matter (domain) expertise |
| | General safety (human factors) experience |
| Control | Subject matter (domain) expertise |
| | Priorities of the organisation (business strategy) |
| Time | Stakeholder-defined deadline |
| | Investigation/analysis schedule update-3 |

# Index

human performance 7, 8, 24, 25, 30, 57, 58, 61, 65, 66, 93
Hume, D. 14

Ignorance 1–4, 9, 12, 13, 16, 92
imagination 3, 10, 13, 36, 37, 90

Kierkegaard, S. 3, 10
Kirwan, B. 10

Latent condition 14, 35, 88
Leucippus of Miletus 12
Leveson, N. 134
Lewes, G. H. 25
linear thinking
    complex 13, 18
    simple 12–14, 18
lines and arrows 43, 44, 55
Lundberg, J. 9, 94

Mach, E. 23
maintenance 2, 35, 65, 68, 69, 74, 109
    inadequate 65, 74
malfunction 7, 9, 12, 22, 31–33, 65, 72, 85, 89, 114, 130, 131
Maruyama, M. 17, 19
method-sine-model 21, 132
model-cum-method 21, 128, 132
monitoring 90, 91, 121, 138

Nietzsche, F. W. 12, 19
noise 7, 28–30, 32
non-causal 15, 29
non-compliance 34, 38, 39
non-linear 18, 25, 28, 31, 35, 85, 88, 89, 132

Operating conditions 68, 74
organisational performance 24, 28, 30, 58, 66, 69, 93
outcomes
    emergent 22, 25, 26, 29, 88, 89, 130
    resultant 22, 25, 89
    unexpected 28, 36, 63, 85
Output
    acceptable 70, 71
    imprecise 70, 71
    precise 71

timing 11, 52, 58, 69–71, 73, 78–84, 92, 124

Performance
    adjustments 31, 32, 38, 85
    conditions 7, 57, 58
    indicators 90, 138
    PSF (Performance Shaping Factor) 7, 57, 58
    variability 24–25, 28, 29, 35–37, 53, 56, 63, 65–69, 71, 77, 85, 87–89, 91, 92, 105, 106, 121, 124, 137
        amplitude 27, 66, 68, 69
        frequency 30, 66, 68, 69, 114
Petroski, H. 1, 2, 9
phenotypes 69, 71
Praetorius, G. 125
precision 69–71, 79–84, 92, 106
prevention 26, 89, 90, 94
principle of decomposition 5, 6, 42
Pringle, J. W. S. 10
probabilities 8, 35, 37, 93, 94
protection 26, 87, 89, 90
PSF (Performance Shaping Factor) 57, 58

Quantification 93, 94

Rasmussen, J. 134
Reason, J. 134
resilience engineering 9, 19, 28, 87, 90, 93, 124, 125
resonance 15, 22, 27–32, 35, 37, 64, 75, 77, 85, 87–89, 134, 137
    classical 27, 28
    functional 22, 27–31, 37, 64, 77, 85, 87, 88, 134
    stochastic 27–32
risk assessment 2, 3, 5, 9, 13, 14, 21, 26, 33–40, 55, 56, 58, 93, 95, 119–123, 127, 134, 136
Ross, D. T. 61

Scale invariance 133
Schulzinger, M. S. 13, 19
second cybernetics
    deviation-amplification 17, 19
    self-regulation 16, 18

Printed in the United States
by Baker & Taylor Publisher Services